Acknowledgements

I would like to sincerely thank the staff of the Wellcome Cell Culture Unit of the Department of Biochemistry at Glasgow University for all the cheerful help they have given for many years. I would also like to thank the Medical Illustration Unit for the illustrations in this book and the secretaries of the Biochemistry Department for typing my, at times scarcely legible, notes. Thanks are also due to all those who have given permission for their figures to be used and to Professor Smellie for support over the years and also to Dr. John Pitts for help in taking several photomicrographs and in providing much valuable information for this book.

*Cell culture for
biochemists*

LABORATORY TECHNIQUES IN BIOCHEMISTRY AND MOLECULAR BIOLOGY

Edited by

T.S. WORK – *'East Lepe', 60 Solent View Road, Cowes, Isle of Wight; formerly N.I.M.R., Mill Hill, London*
R.H. BURDON – *Department of Biochemistry, University of Glasgow, Scotland, U.K.*

ELSEVIER/NORTH-HOLLAND BIOMEDICAL PRESS
AMSTERDAM · NEW YORK · OXFORD

CELL CULTURE
FOR BIOCHEMISTS

R.L.P. Adams

Department of Biochemistry,
University of Glasgow,
Glasgow G12 8QQ,
Scotland, U.K.

1980

ELSEVIER/NORTH-HOLLAND BIOMEDICAL PRESS
AMSTERDAM · NEW YORK · OXFORD

ISBN –series: 0 7204 4200 1
   vol. 8, paperback: 0 444 80199 5
       hardbound: 0 444 80248 7

Published by:

ELSEVIER/NORTH-HOLLAND BIOMEDICAL PRESS
335 JAN VAN GALENSTRAAT, P.O. BOX 211
AMSTERDAM, THE NETHERLANDS

Sole distributors for the U.S.A. and Canada:

ELSEVIER/NORTH-HOLLAND INC.
52 VANDERBILT AVENUE
NEW YORK, N.Y. 10017
U.S.A.

Library of Congress Cataloging in Publication Data
Adams, Roger Lionel Poulter.
   Cell culture for biochemists.

   (Laboratory techniques in biochemistry and
molecular biology; v. 8)
   Bibliography: p.
   Includes index.
   1. Cell culture. 2. Cytology–Laboratory
manuals. I. Title. II. Series. [DNLM:
1. Cells, Cultured. 2. Tissue culture.
W1 LA232K v. 8/QS525 A216e]
QH585.A28   574.87'028   80-11231
ISBN 0-444-80199-5

Printed in The Netherlands

Contents

4

8

CHAPTER 1

Introduction

1.1. Background

Although a number of books are available giving a great deal of information about various aspects of cell culture (Harris, 1964; Willmer, 1965; Habel and Salzman, 1969; Vago, 1971; Whitaker, 1972; Fogh, 1973; Sato, 1973; Paul, 1975; Kuchler, 1977) this book is designed rather for the biochemist or molecular biologist, whose interest in cell culture extends only as far as this technique provides him with material with which he may perform biochemical experiments.

Before a biochemist will apply himself to using the technique of cell culture he must be assured that it offers him significant advantages which outweigh any disadvantages. Furthermore, he must not imagine that the methods are too laborious for routine use or that some degree of black magic is required before success can be achieved. To some extent such fears are based on the experience of workers in the field up to about 1960. In the preceding thirty years nearly all major cell types had been cultivated for varying periods and much descriptive information obtained but this was only as a result of constant dedicated effort.

Since 1960 many of the obstacles have been removed from the path of the biochemist as a result of three advances. Perhaps the most important is that commercial companies now supply media, sera, cells and culture vessels which enable cells to be cultured occasionally or routinely on a scale varying from a growth surface of less than one square centimeter up to several square metres. This service is available only as a result of the description of simple media in which the cells grow well and the development of simple methods for isolation of primary cells, selection of clones and storage of cell lines. The other major fear is one of cost. On a weight for weight basis cultured cells are several orders of magnitude more expensive than, say, rat liver. Thus a rat costs under £2 and yields about 10 g of liver. 10^6 cells obtained from a commercial supplier cost about £1 and so 10 g of cells (2×10^{10} cells) could cost £20,000 although significant price reductions are obtained when buying in quantity. One can produce the same number of cells in one's own laboratory for less (the cost of medium may be as little as £15), but this hides the cost of overheads. Nevertheless, the use of cultured cells must offer marked advantages before it is worthwhile to embark on large scale production, and there is seldom any justification for using cultured cells as a source of material for an enzyme purification when sources like rat liver or rabbit kidney would do as well. From the other point of view many experiments may be performed with 10^2–10^6 cultured cells at a cost equivalent to the alternatives.

1.2. Some advantages

One major advantage offered by cultured cells which cell biologists make full use of but which is often ignored by biochemists is that the living cells may be watched under the microscope. It is essential that healthy cells are used in an experiment and that they remain alive throughout the experiment. That this is the case may be monitored regularly and moreover quantitative estimates of the proportion of viable cells are readily obtained. It is often impossible to know the state of an animal's kidney until the end of the experiment and then usually only in a qualitative manner.

Cells in culture offer a homogeneous population of cells of virtually identical genetic make-up growing in a constant environment. Moreover, the environment may be changed, within limits, at the whim of the experimenter who may thereby investigate the effect of pH, temperature, amino acid and vitamin concentration etc. on the growth of the cells. Growth may be measured over a short time period either by measuring an increase in cell number or size, or by following the incorporation of a radioactive tracer into DNA. These are real advantages over a whole animal system, placing cultured cells on a par with microorganisms as an experimental system. Using cultured cells, the growth requirements of human cells were analysed in a few weeks thus confirming decades of work with people of different genetic background living in different environments (Eagle, 1955a, b; see § 2.2).

Moreover, significant results may be obtained with very few cells. An experiment which may require 100 rats or 1000 humans in order to clarify some point may be statistically equally valid if 100 coverslip cultures are used. If each cell is regarded as an independent experiment then one coverslip culture may yield more reliable results than a hospital full of people. This is obviously a major advantage as far as man is concerned but also overcomes the ethical problems which often arise when large numbers of animals are used for experimental purposes. However, in the final analysis, many experiments must be performed on whole animals, but this is no justification for not using cultured cells for the preliminary work.

Because cells in culture are easily available for manipulation by the biochemist, radioactive tracers, drugs or hormones etc. may be applied in a known concentration and for a known time period. The amounts of such compounds required may be an order of magnitude less than with comparable experiments on whole animals. There is no fear that the drug whose effect is to be investigated is being metabolised by the liver, stored in the muscles and excreted by the kidney. It is usually a simple matter to establish that a substance added to a cell culture remains in contact with the cells in unchanged form at a known concentration for a given time. This enables experiments to yield realistic figures for the rates of incorporation or metabolism of compounds. Such experiments are not without hazards in cultured cells (see Chapter 12) but are very difficult to interpret in whole animals. However, when the aim of the experiment is to find the effect of a drug or cosmetic on an animal, factors which are problems to one biochemist may be the essence of the experiment to another.

1.3. Applications

Cultured cells have given us great insight into the phenomena of cell growth and differentiation and the general characteristics of the growth of cultured cells are discussed in Chapter 2. It should be clear, however, from reading later chapters that, although the detailed nutritional requirements and growth control mechanisms are complex, it is now a simple matter to culture cells in small or large quantities in order to perform biochemical experiments.

1.3.1. Differentiation

The study of differentiation in higher eukaryotes is extremely difficult, but a number of systems are now available which undergo differentiation in vitro and some of these are considered in Chapter 15. The in vitro systems have the advantage that, following a given stimulus, a population of cells will undergo a change which can be easily recognised and quantitatively monitored. The change may be the production of a protein (e.g. haemoglobin by the Friend cells – § 15.1) or more complex alterations in structure and growth pattern such as those occurring during differentiation and fusion of myoblasts (§ 15.5) or differentiation of epidermal keratinocytes to form a system resembling the stratum corneum of skin (§ 15.2).

A large number of cell strains have been derived from the central nervous system including differentiated glial cell lines (Benda et al., 1968; Lightbody et al., 1970; Ponten, 1973) and neuronal cell lines (Augusti-Tocco and Sato, 1969). In addition, Pfeiffer and Wechsler (1972) have isolated clones from tumours of the peripheral nervous system (neoplastic Schwann cells). A book (Sato, 1973) has been written about the special techniques required for the culture of cells of nervous origin. Neuroblastoma cells in culture will extend neurites and, when cultured with differentiating myoblasts, synapse formation occurs allowing signal transmission to be studied in a very simple system (Schubert et al., 1973). Astrocytes and astrocytoma cells may be cultivated in vitro (Guner et al., 1977) where they are readily distinguished from any contaminating fibroblasts by their very long processes which form a reticular network. At present, work is in progress to determine whether they retain the ability to synthesise and degrade neurotransmitters (P.T.F. Vaughan, personal communication). It is always a problem when using specialised cells in culture that some of their characteristic properties may have been lost on isolation, though Pfeiffer and Wechsler (1972) are studying myelin formation and demyelinating disease in cells derived from Schwann cell tumours.

One particular group of tumour cells, the plasmacytomas, are derived from an immunoglobulin producing plasma cell. These cells which may be propagated in vivo or in vitro have enabled rapid advances to be made into the understanding of antibody production. They are being used in the study of immunoglobulin structure and function and in particular the structure and production of messenger RNA including playing a major role in our understanding of RNA processing (Hozumi and Tonegawa, 1976; Weigert et al., 1978); and the unique rearrangements of the antibody genes which result in antibody diversity are playing a role in our understanding of the complex process of differentiation (§ 15.4; Rabbitts and Milstein, 1977).

1.3.2. Genetics

One of the great advantages the bacteriologist had over the traditional biochemist working with eukaryotes was the possession of a wide range of mutants which allowed him to perform complex experiments in genetics. The study of familial relationships in eukaryotes was a time consuming occupation and generation times are particularly long for mammals. This problem is accentuated when human genetics is being studied making it at best only an observational science. With the ability to grow cells in culture came techniques enabling cells to be cloned (§ 8.1), stored (§ 8.3) and fused (§ 13.5) which have led to the science of somatic cell genetics. Many of these studies have centred on a gene whose product, hypoxanthine phosphoribosyl transferase (HPRT), is involved in purine nucleotide biosynthesis and a deficiency of which results in gouty arthritis. Cells with a defective HPRT are unable to incorporate the hypoxanthine analogue 8-azaguanine and so are resistant to this analogue's toxic properties, and cell biologists have exploited the selective advantage of such HPRT⁻ cells in developing techniques of cell hybridisation (§ 13.5) and gene transfer (Goss and Harris, 1975; Willicke et al., 1976a, b).

Human pedigree studies of families with Lesch–Nyhan syndrome (Seegmiller et al., 1976) showed that the HPRT gene was X-linked, and this was confirmed by analysis of human–mouse cell hybrids which had lost most of their human chromosomes (Ricciuti and Ruddle, 1973).

Rather than rely on the random loss of chromosomes from inter-specific hybrids, a number of groups are attempting to introduce into recipient cells single chromosomes either entrapped in phospholipid vesicles (Mukherjee et al., 1978) or by treatment with dimethylsulphoxide (Miller and Ruddle, 1978). The central role of the HPRT locus has been recently reviewed by Caskey and Kruh (1979).

1.3.3. Immunology

Although cell culture has been used by immunologists as described above, for a number of years the existing system was beset by a problem. Cells which were synthesising antibodies of interest (e.g. spleen cells from animals injected with specific antibodies) grow poorly or not at all in culture while the myeloma cells produce an antibody of unknown specificity (§ 15.4). The ability to fuse these two types of cell has recently led to the production of monoclonal antibodies on a large scale (Kohler and Milstein, 1975). If a mouse is injected with a crude preparation of an antigen and its spleen cells are subsequently fused to myeloma cells then among the resulting hybrid cells will be one producing a single antibody directed specifically against the antigen. This cell may be cloned (§ 8.1) and grown as a tumour in a mouse and in this form will yield gram quantities of highly specific antibody. Not only is this a help to the immunologist, it also provides the biochemist with antibodies to material he cannot purify, and it has been used in genetic analysis of human cell surface antigens (Barnstable et al., 1978).

1.3.4. Hormones

The sixth Cold Spring Harbor Meeting on Cell Proliferation (Aug.–Sept. 1978) was concerned with hormones and cell culture (Rudland, 1978). This was divided into the study of growth promoting factors, which is considered in more detail in § 2.2 and § 7.7, and the action of hormones on cellular differentiation. Perhaps falling between the two fields is the work of Guner et al. (1977) which attempts to explain the ameliorating effects of administration of glucocorticoid steroids on patients with brain tumours. They have shown that dexamethasone at low concentrations raises the plating efficiency of human astrocytoma cells and this may be the result of the appearance of a cell surface antigen similar to that observed by Ballard and Tomkins (1969) on treatment of hepatoma tissue culture (HTC) cells with dexamethasone. Cultures of astrocytoma cells, however, achieve a lower terminal cell density and fewer cells make DNA in dense cultures in the presence of dexamethasone (Freshney, 1979); this may be the rationale for the use of glucocorticoids in therapy. Glucocorticoids are also taken up by neoplastic lymphoid cells in cell culture where they cause cell death, and mutants are being isolated to investigate the receptor gene (Bourgeois and Newby, 1977).

The culture of a biopsy of human mammary tumour tissue with a view to determining whether such a tumour would respond to hormone therapy is in its infancy (Leake, 1978). However, Lippman et al. (1977) have obtained a human breast cancer cell line (the MCF-7) which responds to oestrogen treatment by increased uptake of amino acids and nucleosides. It is important to perform the hormone treatment in the absence of interfering substances present in serum; but the use of an in vitro system has enabled the oestrogenic activities of various oestrogens to be compared with their binding affinities for the receptor protein.

The biochemical mechanism of ovulation proposes that plasminogen activator (a serine protease) is responsible for disruption of the graafian follicle and this is being studied in vitro by exposure of ovarian granulosa cell cultures to follicle stimulating hormone and luteinising hormone (Strickland and Beers, 1976).

1.3.5. Production of cell products

The cultured cell may become a major source of hormones and other secreted materials when further work is performed to optimise the system to this end. They are already very important in the production of the species specific antiviral agent interferon. Ogburn et al. (1973) describe a method for isolation of interferon from the culture fluid of mouse L cells exposed to inactivated virus or the non-viral inducer poly I : C.

Interferon is produced in virally infected cells and induces in other cells an antiviral state resulting at least in part from a specific inhibition of viral mRNA translation (Revel and Guner, 1978). Interferon shows host cell specificity and therefore production of human interferon must be performed with human cells. Fibroblasts and leukocytes have been used and the inducing viruses include Sendai and Newcastle Disease Virus (Gresser, 1961; Baron and Isaacs, 1962; Merigan et al., 1966). In addition, homologous cell cultures must be used in the assay of interferon and primary late embryo cultures or foreskin fibroblasts have been used (Merigan et al., 1966). The diploid cell line W138 is less sensitive.

The assay for interferon involves incubating cells overnight with increasing dilutions of interferon and then challenging the cells with, say, vesicular stomatitis virus (VSV) at 20 p.f.u. per cell. Twenty hours later the culture fluids are harvested and assayed for VSV using a plaque assay (§ 14.3.2) on mouse cells. The greatest dilution of interferon which inhibits virus yield by 3.2 fold (0.5 $\log_{10}$) contains 1 unit of interferon (Baron, 1969).

The most suitable system for collection of a cell metabolite secreted into the medium is if the cells can be maintained in capillary beds (§ 3.2.3.4) and Odell et al. (1967), Knazek and Gullino (1973) and Knazek et al. (1974) describe the isolation of chorionic gonadotropin from the perfusate of human choriocarcinoma cells maintained in this manner.

Erythroprotein is present in only very low levels in serum or urine and the idea of obtaining larger amounts from cell culture seems reasonable but has so far met with only limited success. The possible reasons for this have been reviewed recently (Ogle et al., 1978).

1.3.6. Virology and cell transformation

Much of the rapid progress in the field of virology over the past decades is a consequence of the ability to grow viruses in cells in culture. This not only means that large numbers of animals are no longer required but that assays and procedures which were cumbersome and of poor reproducibility have been replaced with plaque assays, production and staining techniques (Chapter 14) which are simple, accurate and reproducible. This has led to the realisation that viruses not only infect and kill cells but may also bring about the change in cell growth characteristics known as viral transformation of cells (§ 2.1 and § 14.4). These changes, which result in the cell no longer responding to its neighbours in a manner characteristic of the untransformed cell, are being studied in order to throw light on the nature of the transformation event as a similar change in vivo is believed to play a part in the induction of tumours (§ 14.4).

As most viral diseases can now be treated by administration of antisera it is important to be able to grow batches of virus both for identification purposes (§ 14.3) and for use in the production of vaccines. Much of this work was done in cultured cells and many hospital virology units are well equipped for growing cells and cultivating viruses on a large scale.

1.3.7. Cytotoxicity testing

A major application foreseen in the use of cell cultures is to test and investigate the mode of action of various products which may be used as drugs, detergents, cosmetics, insecticides and preservatives, etc. Although results obtained using cells in culture cannot be extrapolated directly to the whole animal situation it is fairly certain that if some product produces deleterious effects on several different lines of cells in culture some ill effects may be expected if the product is applied to whole animals. As well as enabling testing to be performed without the possible suffering of large numbers of animals the use of human cells allows testing in one animal species not generally available for experimentation, i.e. man. Moreover, as indicated in § 1.2, the results of the test are more likely to be reproducible when carried out in vitro.

A somewhat different sort of cytotoxicity testing involves the screening of lymphocyte preparations for their antitumour activity. Hellström (1967) describes an assay whereby animals were immunised with tumour cells and the resulting lymphocytes screened for their ability to block the growth of the original tumour cells. Such microtoxicity testing may be performed in volumes of 10 μl to 5 ml and upwards but is most conveniently performed in the 0.3 ml wells of a microtitration plate (Table 3.1) as described by Hellström and Hellström (1971).

Cytotoxicity may be assayed by the measurement of viable cell number (§ 8.2.4), but this is a tedious method to use for large numbers of assays as it requires long hours looking down a microscope. A simpler method which has recently been automated by the introduction of the Titertek supernatant harvester (see Appendix 3) involves the measurement of radioactive chromium released into the culture medium from killed cells. The harvester consists of a set of absorbant cylinders aligned so that they may be inserted into the wells of the Titertek/Linbro microtitration plates (Appendix 3). Once the supernatant in the wells has been absorbed the cylinders are transferred to counting vials and the amount of radioactive chromium released from the cell monolayer is estimated. Cells take up ^{51}Cr sodium chromate rapidly and the excess is readily washed away by rinsing in culture medium. Labelling need only be for 30 min and on subsequent death of the cell more than 75% of the radioactivity is released into the supernatant (Wigzell, 1965; Hirschberg et al., 1977). Although designed for toxicity testing using release of ^{51}Cr the supernatant harvester may well find other uses in biochemistry.

1.4. Plant tissue culture

This is, perhaps, a subject in its own right and the reader is recommended to read the second edition of 'Plant Tissue and Cell Culture' (Street, 1977). Although the principles of the subject are the same as for culture of animal cells a somewhat different methodology has been developed. This is probably partly historical but also partly as a consequence of the differences in the nature of plant cells growing in culture. Thus carrot or tobacco cells may grow as a callus, a suspension, or be induced to undergo embryogenesis (Street, 1975a) or organogenesis (Street, 1975b).

Callus cultures are initiated from shoots or stems. The former are obtained from sterilised seeds induced to germinate under aseptic conditions (Street 1975a) and the latter may be sterilised and then aseptically split to expose the pith to the medium (Street, 1975b).

The culture medium used is generally a modification of that described by Murashige and Skoog (1962) and to this are added the plant growth hormones, 3-indoleacetic acid (IAA) and Kinetin or an analogue thereof (Dalton and Street, 1976). There is, however, a considerable variation between different media in routine use (Heller, 1953; Nitsch and Nitsch, 1956).

Callus cultures are those initially grown from the plant tissue and this sort of culture is used for maintenance of material. Callus cultures are usually maintained on a solid agar medium from which they stand out. For this reason different regions of the callus are in different environments and for most biochemical purposes the callus is dispersed and the cells are grown in suspension. When returned to a solid medium cells undergo an organised sequence of cell divisions giving rise to adventive embryos (Reiner et al., 1977). Recently, plant tissue culture has assumed importance in the horticultural industry. This stems from the finding that from the single cells, separated from a callus culture, individual carrot plants may be grown, i.e. thousands of identical carrot plants may be produced from a single piece of callus. Thus once a plant of improved performance has been obtained there is no need to delay for several years whilst breeding experiments continue and stocks are built up. Rather, within a matter of months, many identical cloned plants are available. Such plants are uniform in colour, size and appearance and uncontrolled variations seldom arise. Gibco and Gibco Biocult (Appendix 3) provide a series of media and media supplements for plant tissue culture and run courses for horticulturalists both in the U.K. and U.S.A.

CHAPTER 2

Characteristics of cultured cells

2.1. Primary cells and transformation

Cells taken from an animal and placed in culture are termed *primary cells* until they are subcultured (Chapter 5). Primary cells, if successfully established in culture, will multiply and will require regular sub-culturing.

On reaching confluence a primary culture may be subcultured into 2 or 4 new bottles and this subculturing may be repeated at about weekly intervals for several months. In such a culture the cells may remain diploid and retain many characteristics of the initial explant. This is a *cell line*, and several different kinds of cells may be present and some of the characteristics may prove unstable. The cells may be cloned (Chapter 8) and some clones may exhibit a stable phenotype. Such is the WI 38 strain, a commercially available strain of human embryonic lung cells, many identical cultures of which were frozen after only a few passages.

Early investigators believed that somatic cells would proliferate indefinitely in culture in an unmodified form if they could be maintained in suitable conditions. However, it is now realised that this is not so. Primary cells are readily established from many tissues and for a while these will proliferate exponentially, but after about 6 months the growth rate falls and by 10 months the cells degenerate and die. This takes place after some 50 generations when about 10^{22} cells have been produced in culture from each initial primary cell (Hayflick and Moorhead, 1961). (Although primary cultures derived from embryonic cells will grow for about 50 generations, cells from adult tissue usually enter senescence after about 20 generations.) During the early stages the cells remain euploid (i.e. have the correct diploid complement of chromosomes) but later they become aneuploid. Very occasionally one of these aneuploid cells will survive and continue to grow, and it is this phenomenon which has led to the origin of the established *cell strains*. The frequency of this transformation event can be increased by treatment of cells with mutagens – e.g. methylcholanthrene treatment led to the isolation of the L strain of mouse cells (Earle, 1943) from which L929 cells were later cloned – or with some viruses (see Chapter 14). The chromosomal pattern of these cells is markedly aneuploid and variable. Thus the chromosome complement of L929 cells varies between 56 and 241 with a mode of 66. Morphological alterations also occur and the L929 lines show no resemblance to the initial fibroblasts from which they are derived.

A cell strain exhibits specific properties which persist during continuous cultivation. Many established cell strains have the capacity to grow and divide indefinitely in culture and in general this is associated with an aneuploid karyotype. Some cell strains, e.g. BHK 21 hamster fibroblasts (Macpherson and Stoker, 1962; Stoker and Macpherson, 1964), have the correct diploid chromosome number but are believed to have the incorrect chromosome complement. These cells, however, are only minimally transformed in that they still exhibit a certain amount of contact inhibition of movement (see § 2.3.3) and they may be further transformed by treatment with polyoma virus (e.g. to form PyY cells) or SV40 (to form SV28 cells) (see Chapter 14).

In contrast, some cell strains, e.g. HeLa (Gey, 1955), that were derived originally from tumour tissues appear to have been transformed in vivo. Although not all neoplastic cell populations will grow indefinitely in vitro many of the human cell lines in common use have such an origin (e.g. HEP 2, KB, Detroit 6).

It is important to realise that there is an alternative explanation to the appearance in a primary culture of rapidly growing 'transformed' cells of different karyotype and morphology, i.e. the primary cells may have become contaminated by a cell line being carried in the same laboratory. This happened in Parker's laboratory when L929 cells contaminated a series of primary cultures and subsequently outgrew the primary cells. Only after careful karyotypic analysis and transplantation specificity tests was this confirmed (Parker, 1959). Concern over a repeat occurrence has led to stringent working routines, but cross-contamination may still occur especially in laboratories (even commercial laboratories) where many cell lines are routinely passaged. Lavappa (1978) has shown that 21 cell lines held by the American Type Culture Collection are derivatives of HeLa. Standard immunological and virus susceptibility tests are now available for species identification.

There is considerable confusion in the literature between the terms cell line and cell strain and many authors use them interchangeably. Others, e.g. Hayflick and Moorhead (1961) define a cell strain as a population of cells derived from a primary culture by subcultivation and a cell line as a population of cells derived from a primary culture and grown indefinitely in vitro.

The relationship between generation number and passage number depends on the *split ratio*. Thus if cells are split so that the contents of one bottle are distributed between two new bottles (split ratio of 1 to 2) then passage number and generation are the same. This is because the cells will only be able to double in number before they again achieve confluence and require subculturing. However, with a split ratio of 1 to 4 the age of the cells in generations will be twice the passage number.

2.2. Nutritional requirements

In the 1940's and early 1950's most cells were grown in plasma or fibrinogen clots, in the presence of tissue extracts and their ultra-filtrates. Two cell lines, the mouse L cell (Sanford et al., 1948) and the HeLa cell (Gey, 1955), were cultured on the surface of glass containers, and in a classical paper Eagle (1955) investigated their nutritional requirements. He was able to propagate these cell lines in the presence of a defined mixture of amino acids, vitamins, salts and carbohydrate supplemented with a small amount of dialysed horse or human serum. Specific nutritional deficiencies were produced by omission from the medium of particular amino acids or vitamins and these could be 'cured' by restoration of the missing component. Twenty-seven factors were defined as essential for growth and they formed the basis of a medium known as 'basal medium, Eagle' or BME. Thirteen amino acids were essential, the remaining six non-essential amino acids being synthesised from other carbon sources. Omission of any one of seven vitamins led to the development of deficiency symptoms. Thus Eagle by using cell culture techniques was able to demonstrate the nutritional requirements of mouse and human cells. This basal medium required frequent replenishment for the cells to continue growing and was shortly replaced by Eagle's minimum essential medium (MEM) (Chapter 7 and Appendix 1) in which the concentrations of the various components are increased to enable cells to continue growing in culture for several days between medium changes.

The undefined factor in Eagle's media is the serum component. There is still considerable controversy over the role played by serum. Is it simply a source of low molecular weight compounds which may be required in very small amounts and which may be carried by serum proteins or is it providing essential proteins or perhaps it provides both large and small molecules? Both Fisher et al. (1958) and Lieberman and Ove (1957) showed that an α-globulin fraction (fetuin) present in adult and foetal serum promoted the attachment of cells to glass and their flattening or spreading, both of which were essential to cell multiplication. The action of fetuin was attributed to its antitryptic activity (Fisher et al., 1958) and today following trypsinisation of tissues (Chapter 6) or cell monolayers (Chapter 5) the continued action of trypsin is routinely blocked by addition of medium supplemented with serum.

It is, however, the property of serum albumin of acting as a carrier for small molecules which has created much interest in the 1970s. These small molecules (hormones) may act both in vivo and in vitro to stimulate growth of different kinds of cells. They are discussed more detailed in § 7.7.

One of the problems that arose when attempts were made to grow single cells in culture or cells at low density (e.g. 100 cells/ml) was that all of the cells died or grew only very slowly. Cells were able to grow at low density if, in addition to the normal requirements, serine (Lockart and Eagle, 1959) or cystine (Eagle et al., 1961) were supplied. These and other population dependent requirements were investigated by Eagle and Piez (1962) and this led to the concept of 'conditioned medium'. Although cells are able to synthesise these additional requirements they are lost from the cell to the environment in amounts which exceed the biosynthetic capacity of the cell. Conditioned medium is medium in which the concentration of metabolites has built up to such a level that an equilibrium is achieved between metabolites lost from the cell to the medium and metabolites taken up from the medium by the cells.

2.3. Growth control

2.3.1. Cell cycle and growth cycle

Growing cells undergo regular divisions about once every 24 h. In between divisions (i.e. during interphase) they double their complement of DNA during a distinct period known as the DNA synthetic or S-phase. S-phase is separated from cell division or mitosis (M) by two gaps (G1 and G2) (see Fig. 10.1). Cells which are not restricted in any way will proceed indefinitely around this cycle (the cell cycle) and are said to be in exponential growth as the cell number doubles on each circuit.

Normally, after a short period of exponential growth, some factor becomes limiting. Possibly the area for growth is completely covered or some factor in the medium becomes exhausted (see § 2.3.3). The rate of growth of the culture then slows down and the size of the culture reaches a plateau (the terminal cell density). When no further cell division occurs the cells are in a stationary phase. When the limiting factor is restored such cells reenter the cell cycle in the G1 phase and undergo a round of DNA synthesis prior to cell division. There are two or more theories which attempt to explain this type of growth control (§ 10.4) but they all propose a point of control shortly after cell division. Once this point has been passed a cell will proceed around the cell cycle and divide.

As most of the cells in an animal are under some form of growth control they are not proceeding around the cell cycle but have been arrested shortly after mitosis. Thus on setting up a culture of primary cells the restrictions on growth must be removed before those cells can proceed towards division. The cells in some rapidly growing tumours appear to have lost their susceptibility to growth control and for this reason primary cells obtained from tumours may be readily established in culture and may grow to higher densities than cells from normal tissues (see below).

Giant cells sometimes arise in cultures especially if the growth conditions are not optimal. They are produced by the failure of growing cells to divide and they may increase in size until they are 1 mm or more in diameter. The incidence of such cells is markedly increased by irradiation (Tolmach and Marcus, 1960). The presence of an occasional relatively small giant cell in a population probably represents no threat to biochemical experimentation, but if their incidence rises it is a reflection of poor culture conditions; such cultures should be discarded and fresh ones obtained and grown in improved media.

2.3.2. Anchorage dependence and growth in suspension

Although lymphocytes show no tendency to aggregate in vivo and will grow in suspension in vitro (§ 6.3) most untransformed mammalian cells both in vivo and in vitro grow attached to a substratum either of other cells, collagen, or of glass or plastic (Klebe, 1974). Plastic surfaces need to be specially treated before cells will attach and eukaryotic cells will not attach to bacterial plastic dishes. When buying plastic ware it is important to specify that it is for tissue culture; the letters TC frequently appear in the catalogue number.

Collagen is the natural substratum on which fibroblasts grow in vivo and Klebe (1974) describes the preparation of collagen coated dishes. Another favourite substratum for studying anchorage dependence is gelatin (denatured collagen). Dishes may be treated with an aqueous gelatin solution (1% for 2 h at 4°C) and then washed with water and stored at room temperature until required.

Although cells growing attached to a substratum have advantages in some systems, for other purposes a suspension culture may be preferable. In general, however, those cells dislodged from a substratum on which they are growing fail to grow in suspension and quickly degenerate. Earle et al. (1954) showed that if L cells were maintained in a roller bottle rotating at 40 r.p.m. they failed to attach to the surface and that the addition of methylcellulose (Methocel) at 0.1% prevented clumping and maintained viability. A number of cell strains have now been selected which grow readily in suspension, e.g. HeLa S3 and LS cells, while other cell strains, e.g. A9, will grow either attached to the substratum or in suspension depending on the nature of the salt solution in which they are grown, i.e. omission of divalent ions and an increase in the phosphate concentration favours growth of cells in suspension (see § 7.2 and Eagle, 1959).

2.3.3. Density dependent regulation (contact inhibition)

Primary cells will continue or start to divide in culture but exhibit contact inhibition of movement (Abercrombie and Heaysman, 1954). When two such cells approach one another the characteristic ruffling movements of the cell membrane stop in the area of contact. Primary cells therefore do not grow one on top of the other and, in general, cease to divide when a monolayer has been formed. This phenomenon is not restricted to primary cells but applies also to many cell lines. An ideal example is the 3T3 mouse fibroblast cell line which grows rapidly in sparse culture but all division stops as soon as the cells become confluent at about 10^6 cells per 6 cm dish (Holley and Kiernan, 1968). Such non-transformed cells may for some time remain healthy in this quiescent state. The actual cell density is related to the concentration of serum in the medium and Todaro et al. (1965) have shown that addition of serum to an inhibited culture results in a round of DNA synthesis and cell division. A number of factors have been isolated from serum which show some ability to overcome contact inhibition (Holley and Kiernan, 1968; Holley, 1975) and cells transformed by viruses (§ 13.4) show reduced contact inhibition (Holley and Kiernan, 1968; Dulbecco, 1970) and grow to a higher terminal cell density. They are said to have lost density dependent regulation. Transformed cells, in contrast to untransformed cells, continue to grow until they have exhausted the medium, and unless this is quickly replenished such cells soon die. It would seem that the growth of transformed cells is less dependent on the macromolecular components of serum (i.e. hormones or growth factors: § 7.7) and becomes limited only when some of the low molecular weight nutrients become exhausted.

As cells in a monolayer grow two changes occur: 1) the cells become more crowded and less flattened and so expose a diminished surface area to the medium; 2) the medium becomes depleted in nutrients etc., especially in a zone immediately surrounding the cells (Stoker, 1973).

If a strip of cells is removed from a confluent monolayer of untransformed cells (e.g. 3T3 mouse embryo cells) then the cells at the edge of the wound are stimulated to synthesise DNA and divide. They quickly colonise the unoccupied area of the wound. This phenomenon known as topo inhibition (Dulbecco, 1970) is now explained by the presence in cells on the edge of the wound of an increased surface area exposed to the medium (i.e. neighbouring cells have been removed) (Stoker, 1973; Dulbecco and Elkington, 1973).

The fact that crowded cells are less flattened and that cells in suspension are spherical and fail to grow, was followed up by Folkman and Moscona (1978) who showed that the extent of cell spreading was closely related to growth in non-transformed cells. However, whether cells actively control their shape and hence their rate of uptake of nutrients and rate of growth is unclear.

Cell–cell adhesion takes place in three stages described by Walther et al. (1976) as follows:

1. Immediate attachment not requiring energy.
2. After 2 min at 27°C a bond is formed between cells which is susceptible to 0.01% trypsin. This bond is only formed between cells and not with the substratum.
3. After 8 min at 37°C a more stable bond is formed.

However, except at gap junctions (see § 13.6), cells do not approach each other, or the substratum, closer than about 45 nm (Klebe, 1974).

2.3.4. The cell membrane

The cell membrane is a fluid, partly negatively charged, bilayer. Negatively charged glycoproteins protrude on the outside of the membrane but their polypeptide chains pass through the membrane and associate with intracellular proteins. It is these cell surface glycoproteins which play a major role in cell–cell recognition and intercellular adhesion. Thus Abercrombie and Ambrose (1962) first suggested that the cell surface glycoproteins may play a role in contact inhibition (§ 2.3.2) and Aaronson and Todaro (1968) showed that the saturation density of cells in culture correlates closely with tumorogenicity when the cells are injected into mice; i.e. there is a relationship between cell surface glycoproteins and the transformed state. It has been shown that the cell surface glycoproteins serve to mask the agglutination sites on the cell membrane recognised by plant lectins, e.g. phytohaemagglutinin and concanavalin, and this prevents clumping of non-transformed cells by these lectins (Aub et al., 1963). However, agglutination sites are exposed in tumour cells and may be partly exposed in untransformed cells by treatment with 0.005% pronase or 0.007% trypsin. This treatment leads to a partial loss of contact inhibition and elicits a round of cell division in confluent 3T3 cells (Burger, 1970).

Also masked by cell surface glycoproteins are the histocompatibility (H-2) antigens of mouse and this affects the transplantability of ascites cells (Sanford et al., 1973).

The major cell surface glycoprotein is fibronectin (Yamada and Olden, 1978; Olden et al., 1979). This is also known as cig (cold insoluble globulin), CSP (cell surface protein), α2-SB (surface binding glycoprotein) and LETS (large, external, transformation-sensitive) protein. Fibronectin has a molecular weight of 220,000 but exists as disulphide linked dimers or higher oligomers and is found in serum as well as on the surface of normal but not transformed cells. It can be removed from the cell surface by treatment with very low levels of trypsin (Pearlstein, 1976). Fibronectin is not present on mitotic cells and increases in amount on normal cells as they reach confluency or in cells arrested by low serum treatment (§ 11.6) (Pearlstein, 1976; Hynes and Bye, 1974).

Addition of fibronectin to transformed cells causes a partial return to the normal phenotype in that it increases the adhesion of cells to other cells and to the substratum. It is probably one of the factors in serum necessary to promote attachment and spreading of cells in the culture dish. Moreover, antibodies to fibronectin will induce some characteristics of transformed cells in otherwise normal cells.

Fibronectin is considered not to be an integral membrane protein as it may be released from the cell surface with 1 M urea. It is a very flexible molecule consisting of several loosely linked domains (Alexander et al., 1978) and forms a relatively immobile fibrillar network on the cell surface which, in some way, forms a trans-membrane association with cytoskeletal elements (Lazarides and Revel, 1979).

2.4. Differentiated functions in cell cultures

Although in most cases cells derived from primary cultures quickly revert to undifferentiated fibroblasts and epithelial cells, an increasing number of cell lines have been produced which retain some of the functions of the tissue of origin. One of the major problems in establishing lines of differentiated cells was the presence of the much faster growing undifferentiated fibroblasts in the initial explant (see Chapter 15 and Fig. 6.2). Normally these fibroblasts rapidly outgrow the differentiated epithelial cells. In some instances, however, the ready attachment of fibroblasts to a substratum has been used to remove them from a mixed culture (Yaffe, 1968; Rheinwald and Green, 1975).

Puri and Turner (1978) report that in the absence of serum only fibroblasts from chick muscle tissue adhere to the substratum thus leaving the myoblasts in suspension from where they may be recovered and transferred to a second vessel in serum containing medium (§ 15.5).

By first converting a solid hepatoma into an ascites cell line, Thompson et al. (1966) were able to establish lines in culture (hepatoma, tissue culture or HTC) which responded to dexamethasone (a steroid hormone analogue) by induction of tyrosine transaminase, i.e. a typical liver cell response.

Another method of removing fast growing fibroblasts is to prepare primary cultures from a tumour and after a short time return them to an animal where they will reform a tumour. By repeated alternation of growth in vivo and in vitro it was hoped to select for tumour cells and this has proved successful in isolating differentiated cell lines from adrenal, pituitary and neural tissue (Buonassisi et al., 1962; Augusti-Tocco and Sato, 1969).

The loss of differentiated functions is often associated with a high growth rate of the cells in culture and when growth or DNA synthesis is inhibited differentiated functions reappear. Thus a rat pituitary cell line, although normally producing growth hormone, will also synthesise prolactin in response to treatment with bromodeoxyuridine at 3 μg/ml (Biswas et al., 1977) and the Friend cells (§ 15.1) cease to grow and start to make haemoglobin when treated with dimethylsulphoxide or butyric acid.

Some systems where differentiation may be elicited in vitro are considered in Chapter 15.

2.5. *Commercial supplies and transport*

Some primary cells are now available from commercial companies, e.g. Flow Laboratories and Gibco Biocult (Appendix 3), and the methods of preparation are given in Chapter 6. If a company is nearby and will deliver the cells fresh on the day of isolation these may well be a suitable source of material, but otherwise it may be advisable to prepare primary cells oneself. In any case this is essential when the tissue is from an experimental subject.

Cell strains and cell lines are readily available from commercial companies or from the American Type Culture Collection (ATCC), when they come with their pedigree. Obtaining cells from the ATCC can be expensive, and unless one is experienced at growing the cells in question it may be advisable to get a local commercial company to obtain the cells and establish them so that further supplies will continue to be available.

ATCC maintains its stocks as vials of cells frozen in liquid nitrogen (see § 8.3) and they are transported frozen in solid CO_2. However, as cell viability is poor at dry ice temperature, it is important that as soon as the frozen cells arrive they are refrozen in liquid nitrogen or used.

Unfrozen cells may be transported as a monolayer either in a vessel with only a trace of medium or in a vessel completely filled with medium. In either case the bottle must be completely sealed to prevent loss of medium by evaporation and it must not be exposed to temperature outside the range of $10°-37°C$. Even then damage to cells may occur and this method of transport should only be used locally.

Cell culture media and serum are most conveniently bought from a company such as Flow Laboratories or Gibco Biocult (Appendix 3). Medium may be bought which requires only the addition of serum before use, but various concentrates and powdered media are much cheaper to use and are recommended when larger quantities are required. The reconstitution of medium is considered in Chapter 7 and Appendix 1.

CHAPTER 3

Culture vessels

3.1. Design of culture vessels

For the design of culture vessels the following factors are significant:
1. Do the cells grow in suspension or as a monolayer?
2. The scale of the operation, i.e. are single cells the object of study or are gram quantities required? From the biochemists point of view the requirements for their subsequent manipulations must be taken into account.
3. Is gaseous exchange with the atmosphere allowed or should the vessel be sealed off?

3.1.1. Gaseous exchange

Gaseous exchange is possible; it occurs with Petri dishes and cell culture trays and some types of suspension vessels. When the medium used is buffered with bicarbonate (30 mM) it is essential that such cultures be maintained in an atmosphere of about 5% CO_2 in air. This allows maintenance of the correct pH which is readily monitored by the colour of the phenol red present in the medium. The colour should be a tomato colour – not yellow and not red and certainly not purple – indicating a pH of 7.2. Alternatively, media buffered to pH 7.2 with Hepes buffer (20 mM) (see § 7.2.1) can be used when control of the atmospheric pCO_2 is unnecessary. In some instances, especially cell cloning, Hepes buffer (20–25 mM) is used in combination with 8 mM bicarbonate when the vessels should be maintained in equilibrium with 2–3% CO_2.

3.1.2. Sealed vessels

These are, for instance, bottles with bungs or tight screw caps and sealed suspension vessels. If the medium is buffered with Hepes no special precautions are required to maintain pH in the early stages of growth. When bicarbonate is the buffer the pH must be maintained by introducing CO_2 into the vessel prior to sealing. This may be done by passing 5% CO_2 in air through the bottle for about 15 sec or by introducing a fixed volume of CO_2 into the bottle (see Table 3.1 for volumes required, and Fig. 3.1). In both cases the CO_2 is administered through a sterile plugged Pasteur pipette. The mixture of 5% CO_2 in air may be obtained directly from a cylinder available from e.g. British Oxygen Company Ltd., or may be mixed from air and CO_2 (obtainable from Distillers Co., Ltd.). Mixers are built into many CO_2 incubators but are readily constructed from two gas flow meters and an air pump (see Fig. 3.1).

Specially designed CO_2 incubators are available from a number of companies, e.g. Bellco, Forma, Gallenkamp, Grant, Heraeus and Leec (see Appendix).

TABLE 3.1

Container	Surface area (cm^2)	Medium required (ml)	Typical cell volume of	
			Inoculum	100% CO_2[a]
Multiwell or microtitration well	0.3	0.1	$< 1 \times 10^4$	–
TC well	2	0.5	5×10^4	–
5 cm dish	19.4	5	3×10^5	–
9 cm dish	63	10	1×10^6	–
60 ml bottle	16	5	4×10^5	3 ml
125 ml bottle	22	10	6×10^5	6 ml
250 ml bottle	30	15	1×10^6	12 ml
Roux bottle	200	50	3×10^6	50 ml
Thompson bottle	375	100	4×10^6	70 ml
Winchester	1500	150	2×10^7	120 ml
Nunc cell factory	10×600	1500	1×10^8	–

[a] Volume of CO_2 required to adjust the atmosphere to 5% CO_2 in air (see § 2.1.2).

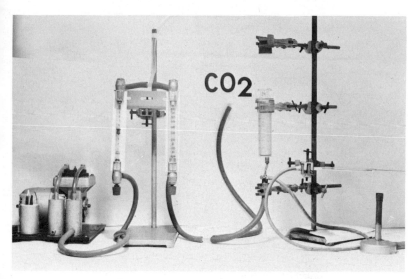

Fig. 3.1. CO_2 delivery systems. On the left is a simple arrangement where air (from an air pump) may be mixed with CO_2 to produce 5% CO_2 in air for gasing bottles or feeding to an incubator. On the right is an arrangement for measuring out fixed volumes of CO_2 and delivering them through a three way tap (at the bottom of the syringe) and a sterile Pasteur pipette to a bottle of cells.

3.1.3. Perfusion techniques

Here gaseous exchange (and medium replenishment) occurs in a second vessel distinct from the cell growth chamber.

3.2. Monolayer cultures

3.2.1. Small scale cultures (Fig. 3.2)

For those interested in cloning reference should be made to Chapter 8.
    A small scale perfusion vessel is considered in § 3.2.3.3.
    For many biochemical studies involving incubation of cells with radioisotopes in the presence of drugs, anti-metabolites, hormones etc. very small numbers of cells are required and these may conveniently be grown on the bottoms of glass scintillation vials or

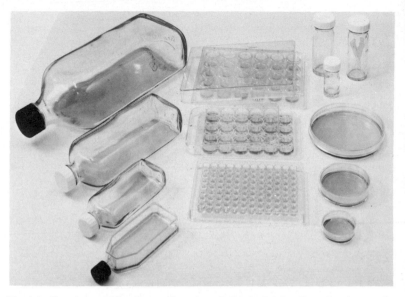

Fig. 3.2. Vessels in which cells may be cultured. On the left is a Roux bottle, two sizes of medical flat bottles (125 ml and 250 ml) and a small Falcon flask. In the centre are two types of trays which hold 24 cover slips in separate wells (Linbro and Falcon) and a 96 well microtitre plate. On the right are three sizes of Petri dish and above them two universal containers and a bijou bottle. In one of the universal containers are three automatic pipette tips ready for sterilisation by autoclaving.

on 1.5 cm diameter glass coverslips or even in the wells of a microtitre plate (see § 1.3.7 and Table 3.1). This last method enables 96 replicate cultures to be handled simultaneously but the maximum volume that each well will hold is 0.25 ml.

Coverslips are used to greatest advantage when placed in the 24 wells of a tissue culture tray. Each well with its coverslip requires 0.5 ml of medium and can be seeded initially with around 20,000 cells.

Scintillation vials require twice this amount and, although more difficult to handle in large quantities, offer the advantage that at the end of the experiment the labelled cells require fewer manipulations prior to estimation of the extent of radioactive incorporation. Alternatively several coverslips may be placed in a 5 cm plastic Petri dish prior to seeding cells.

This has the advantage that each coverslip culture is maintained under identical condition.

Coverslips require very thorough washing before cells will grow on them. They should be dropped individually into boiling 0.1 N NaOH or Chloros, then rinsed overnight in running tap water and afterwards given three rinses in distilled water. They should be laid out to dry on clean tissue and sterilised in glass Petri dishes in batches using dry heat.

The plastic containers are bought in sterile wraps from commercial suppliers (e.g. Nunc, Falcon, Linbro) (see Appendix 3).

3.2.2. Intermediate scale cultures

Cells may be grown in dishes or bottles where the initial inoculum varies from 0.2×10^6 up to 2×10^6. The containers may be glass or plastic. The plastic ware is obtained in sterile wraps from commercial suppliers and is specially prepared for use in cell culture (see Appendix 3). The glass bottles are usually medical flat bottles but any bottle with a flat side will do provided it is washed correctly and sterilised before use (see Chapter 4).

Some glass bottles, e.g. Roux bottles, are manufactured especially for growth of cells in culture and these have the advantage of improved optical qualities allowing easier microscopic examination of the growing cells. Some of the larger bottles are known by the name of the person who introduced them.

3.2.3. Large scale cultures

3.2.3.1. Roller vessels

The simplest of these is the roller bottle. This may be a clear 2 or 2.5 l Winchester bottle or one of the commercially produced roller bottles which have improved optical properties. The cells grow on the inner surface of the bottle (surface area up to 2000 cm^2) and are constantly bathed in medium (about 150 ml/bottle) as the bottle is rotated. Machines to rotate the bottles are available commercially from a number of companies (e.g. Luckham and Voss Instruments; Appendix 3) and many home made ones are in use (see Fig. 3.3). In Glasgow one type of machine made in the Biochemistry Department will roll up to 120 bottles. The speed of rotation of the bottles should not exceed 1 r.p.m. and this is too fast for some cells in the initial

Fig. 3.3. Roller bottle machines. This photograph shows three different models. The one at the rear is the Luckham 6-tier model modified by the manufacturer by provision of a more powerful motor. It is very heavy to manoeuvre and only three standard ($4\frac{1}{2}''$) Winchester bottles can be accommodated on each tier. In the foreground are two tiers of a stackable model made by Voss Instruments. Each tier has a separate motor. On the left is a machine made by Mr. Harvey, Dept. of Biochemistry, University of Glasgow. A motor drives a rubber belt which rotates the bottles as they rest on small pulley wheels. Each motor drives 40 bottles.

stages of growth where rotation at 0.25 or 0.5 r.p.m. improves the attachment of cells.

Sterilin Ltd. (Appendix 3) produce a 2 l polystyrene bottle containing a Melinex (ICI Ltd.) spiral 200 × 20 cm with a 4 mm spacing between the layers. This bottle is filled with 1.6 l growth medium containing 0.4 ml of MS antifoam emulsion RD (Hopkin and Williams) and about 10^8 cells. The bottle is placed on a roller machine overnight to allow the cells to attach and is then removed and gased in an upright position with 5% CO_2 in air. The cells have to be removed by trypsinisation which is a disadvantage if pulse labelling or cell surface properties are under study. The bottle may be taken apart using a hot wire and the spiral sheet with the cells attached removed and the cells scraped off. A 20-fold increase in the number of BHK21C13 cells is obtained within 4 days (House, 1973).

A similar but slightly larger 'Rola Cartridge' is available from Flow Laboratories (Appendix 3). It consists of a 2.27 l container (which may be autoclaved) into which is fitted the sterile cartridge which consists of 40 discs each of surface area 157.4 cm^2. When on a roller machine this requires just over a litre of medium to cover the cells and at the end of growth it may be disassembled for easy removal of the cells. Experience has shown, however, that medium leaks out from the unsealed joins and this type of vessel has to be used with caution.

Abbot Laboratories (Appendix 3) manufacture a Mass Tissue Culture Propagator (MTCP) where the cells grow on glass plates within a large jar through which passes 5% CO_2 in air (Schleicher, 1973) and Connaught Laboratories (Appendix 3) manufacture a similar Multi-Surface Cell propagator. These pieces of apparatus are expensive and difficult to use but can be made to yield cells in 100 g amounts. However, they can be dismantled which give two major advantages.

1. The cells may be removed from the plates by mechanical means.

2. The apparatus is re-usable and, although initially costly, does not require vast continuous expenditure. However, it puts the onus on the user to prepare the apparatus in a clean and sterile manner.

3.2.3.2. Cell factories

Nunc (Gibco-Biocult; Appendix 3) produces a piece of apparatus rather like ten large trays sealed together – the multi-tray unit or cell factory (Fig. 3.4). The surface area for cell growth is 6000 cm^2 and this has the advantage that no rolling is required. The cell suspension (about 1.5 l) is introduced by gravity feed from a bottle in less than a minute and the cells (over 10^9) can be easily harvested by trypsinisation.

Fig. 3.4. A Nunc cell factory. (Courtesy of Nunc and Gibco Biocult.)

It is not possible to remove the cells mechanically unless a way can be developed to dismantle the apparatus using a hot wire. Moreover, the 'factory' is very expensive and can be used once only. However, it produces extremely high cell yields using very little space and in contrast to roller cultures requires no machinery which itself is expensive and liable to break down at critical times. As it is used only once and is supplied in a sterile pack the chances of contamination are slight. Some laboratories have constructed racks to hold several factories so that one person can seed or harvest them simultaneously. When filling, the unit is on its side and the design of the inlet channel ensures that the cell suspension enters all ten chambers equally. The unit is then turned first on end to prevent mixing of medium between the chambers and then placed horizontally. One slight disadvantage with the multitray unit is the large amount of cell suspension initially required to fill it, but this can be overcome by using a single tray unit also supplied by Nunc.

3.2.3.3. Perfusion vessels

Various other procedures have been devised to increase the surface area within a vessel upon which cells may grow. The problem is always to maintain an adequate supply of nutrients including oxygen and to remove waste products, particularly acid. The New Brunswick Scientific Co. (Appendix 3) produce a piece of apparatus which continuously perfuses roller bottles by means of a rotating cap through which the various feed tubes pass.

A small scale perfusion vessel is available as a sterile pack from Sterilin Ltd. (Appendix 3) for microcinematography. The chamber volume is only 0.4 ml but it may be attached to a heated microscope stage and individual cells photographed intermittently. Medium may be circulated around a moat surrounding the culture. It is introduced and leaves through needles let into the sides of the chamber (Cruickshank et al., 1959). The coverslip on which the cells grow may be removed and the cells stained or otherwise processed.

3.2.3.4. Capillary beds

One of the factors limiting the growth of fibroblasts is the availability of medium factors (Dulbecco and Elkington, 1973). Periodic replacement of the nutrient is therefore essential to obtain high cell densities yet has the disadvantage that the composition of the medium bathing the cells is continuously changing. Cell culture on artificial capillary beds does not have this disadvantage but is still not a very popular technique due partly to difficulty in its use and partly to problems with cell harvesting.

Different sorts of capillary beds are available from Amicon Corp. or Bio Rad Laboratories (Appendix 3). These are based on the semipermeable membranes made by these companies for use in their dialysis and concentration cells. Two sorts of capillaries are used: one for exchange of small molecules in solution and the other of silicone polycarbonate for gaseous exchange. The bundle of capillaries (about 150) is held in a tube into which a cell suspension (about 10^6 cells in 3 ml medium) is injected by an opening at the side (Fig. 3.5). When the cells have attached, medium which has been oxygenated and exposed to 5% CO_2 to adjust the pH, is pumped through the capillary bed at a rate of about 1.0 ml/min. The gases will diffuse through the silicone tubing used to carry the perfusion medium.

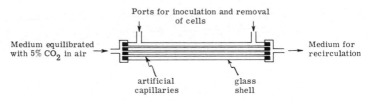

Fig. 3.5. Diagrammatic representation of a capillary perfusion bed.

About 100 ml of perfusion medium is re-used over a period of 2 days before it and the extracapillary medium are changed. The cells remain healthy for a month or more and three small capillary beds in parallel will produce over 2×10^8 cells. This method is extremely useful for the production or metabolic conversion of metabolites. Thus human chorionic gonadotropin may be isolated from the perfusate of human choriocarcinoma cells (Odell et al., 1967; Knazek and Gullino, 1973; Knazek et al., 1974).

3.3. Suspension cultures

Most cells do not readily grow in suspension, and so this method has limited applicability. However, lymphocytes show no tendency to adhere to glass or plastic surfaces and will survive if allowed to sit on the bottom of a tube, dish or bottle covered with a shallow layer of medium. Other cells that are grown in suspension have been selected on the basis that they show poor adhesion to the substratum and that they continue to grow if maintained in suspension. If not continuously agitated they will, however, settle down and adhere to

Fig. 3.6. Spinner flasks. The flask on the stirrer is protected from excessive heat by a layer of foam insulation. The vessel on the right is wrapped in aluminium foil for sterilisation. The model with the solid spindle which is allowed to swivel top and bottom is preferable to the one with metal swivel (on the left) as the latter tends to stir erratically.

the substratum. With some cells it is sufficient to maintain them in roller bottles rotating at about 2 r.p.m., but in general special suspension culture vessels are required and a medium deficient in Mg^{2+} and Ca^{2+} is used.

The suspension culture vessels come in various sizes (from 10 ml to 10 l) and have a magnet suspended just above the bottom of the vessel.

The magnet is driven by a magnetic stirrer motor on which the vessel stands. Most standard magnetic stirrers when running for long periods create excessive heat and the vessel requires to be insulated by having a sheet of expanded polystyrene foam interposed between it and the motor. 'Bellco' spinner flasks (Fig. 3.6) are available from Arnold Horwell Ltd. (Appendix 3).

Suspension cultures grow best within a limited concentration range and when the vessel is about half full of culture medium. In order to maintain these conditions it is necessary to give the cultures regular attention. Every day (or in the case of slower growing cells every other day) half of the suspension should be withdrawn through the side arm and replaced with an equal volume of fresh medium.

3.4. Microcarriers

Microcarriers are small solid particles (kept in suspension by stirring) upon which cells may grow as a monolayer. They confer the advantages of large scale suspension cultures on anchorage dependent cells. The surface of the microcarrier beads should have a positive charge, but different cells show optimum growth at slightly different surface charge. Flow Laboratories (Appendix 3) supply 'superbeads' which they claim have a surface charge suitable for the 'maximum' number of cell types.

The density of the microcarriers should be about 1, so as to facilitate suspension, and Van Wezel (1973) describes the preparation of 100–350 μm diameter beads of DEAE–Sephadex A50.

Flow Laboratories 'superbeads' have a diameter of 150–300 μm and are used at 5 g/l, which makes them an inexpensive substitute for large scale monolayer-culture vessels. The suspension can be sampled and observed readily as the beads themselves are virtually transparent. It is, however, essential to prevent the build up of acid as very high cell densities are achieved (3×10^7 cells in 100 ml suspension will multiply 10–15 fold in 5 days). A combination of Hepes and high bicarbonate is necessary to maintain pH and a better system would involve continuous addition of bicarbonate controlled by pH stat.

The cells may be harvested by allowing the beads to settle and by replacing the medium with 0.25% trypsin in calcium and magnesium free Earle's BSS. This suspension should be stirred at 37°C and monitored microscopically for cell release, which may require vigorous pipetting. The released cells may be separated from the beads using a sintered glass filter funnel or additional beads, and fresh medium may be added and the suspension disposed in fresh vessels for stepwise growth.

David Lewis of Flow Laboratories has grown the following cells on 'superbeads':

Monkey primary kidney
Human embryonic lung
Human embryonic tonsil
Wl 38
Bovine embryonic kidney
Foetal bovine testis
BHK21
CHO-K2
Mouse macrophages

Figure 3.7 shows the rate of growth of BHK21 cells on superbeads and Figure 3.8 shows the appearance of CHO cells growing on superbeads.

'Cytodex 1', recently made available by Pharmacia Ltd. (Appendix 3), are microcarrier beads of Sephadex with a density of 1.03 g/ml and median diameter of 195 μm in PBS. Thus the surface area of 1 g of beads is 0.6 m^2 which is equivalent to 4–6 roller bottles or 100 9–10-cm diameter Petri dishes (see Table 3.1). In theory, therefore,

1 g of 'Cytodex 1' could yield 2.2×10^9 cells. In practice, 25 million cells inoculated into 500 ml of growth medium containing 1–2 g Cytodex 1 would yield 500 million cells in 7 days by which time the medium would be exhausted.

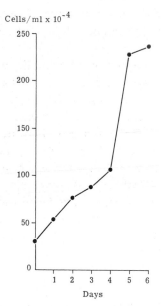

Fig. 3.7. Growth of cells on Flow superbeads. 30×10^6 BHK21/C13 cells were inoculated into 100 ml growth medium containing 0.5 g Flow superbeads and the vessel stirred as for suspension culture. Cell growth was monitored over 6 days. (Reproduced with kind permission of Dr. David Lewis, Flow Laboratories.)

In contrast to Flow's superbeads, Cytodex 1 beads are supplied dry and require swelling in PBS (50 ml/g) and autoclaving before use. In dry form they are stable for more than 2 years.

Pharmacia (1978) report the successful growth of the following cells on Cytodex 1:

Human embryonic lung
Chimpanzee liver fibroblasts

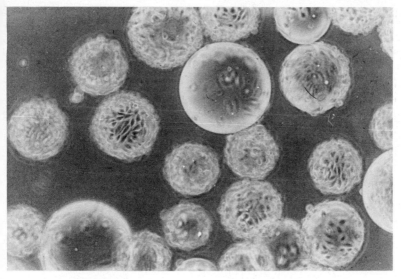

Fig. 3.8. CHO-KI cells growing on Flow superbeads. (Reproduced with kind permission of Dr. David Lewis, Flow Laboratories.)

Human foreskin cells
Chick embryo fibroblasts
Primary monkey kidney cells
BHK 21
HeLa
'Biosilon' beads are microcarriers manufactured by Nunc (Appendix 3). Biosilon consists of particles of specially treated and radiation sterilised plastic material of diameter 160–300 μm. It, therefore, has the advantage of unlimited shelf life and requires no swelling prior to use. 1 kg Biosilon has a culture area of about 25.5 m^2 and a specific gravity of 1.05 and so may be used in the same manner as the other microcarriers.

Bio-Rad Laboratories supply 'Bio-Carriers'; beads which have a cross-linked polyacrylamide matrix. As well as being used in the conventional manner these beads may be added to a slowly rotating roller bottle when they will attach to the glass increasing the surface area many fold. They may be dislodged by shaking and hence easily removed from the bottle.

The use of microcarriers, though still in its infancy, has great potential for industrial scale production of animal cells, especially in the field of virus growth and production by cells of various products which can be isolated from the culture medium obtained simply by allowing the beads plus cells to settle.

CHAPTER 4

Glassware preparation and sterilisation techniques

4.1. General

Cells in culture are fastidious. Not only do they require the absence of any toxic material in their growth media (which must be made up using high quality glass distilled water) but any glassware with which they or their growth media have contact must also be scrupulously prepared. It is largely for this reason that commercially available tissue culture plasticware is so popular as its use, in general, eliminates two, otherwise frequent, suspicions: 1) were the bottles and pipettes clean? and 2) were they sterile? Nonetheless, the extensive use of disposable plasticware is very expensive and most laboratories will find the need to recycle glass vessels.

After use all contaminated glassware should be soaked overnight in baths of disinfectant. Highly recommended is 5% chloros manufactured by ICI Ltd. and available in 5 gallon drums from Durham Chemical Distributors, Birtley, Tyne and Weir or Scottish Agricultural Industries, Renfrew. This is an aqueous solution of sodium hypochlorite and so is not suitable for metal caps and instruments which should be decontaminated by autoclaving. It is very caustic and care must be taken not to splash it onto clothes or exposed areas of skin. Chloros has the effect of killing any viruses, bacteria or cells which may be present, and also prevents proteinaceous materials from drying out and becoming irreversibly fixed to the glass surface.

Uncontaminated glassware should be immersed in baths of water. It is usually convenient to have baths of chloros or water placed strategically in the laboratory and these should be cleared once a day.

The chloros must be renewed weekly as its effectiveness quickly wears off.

After the indicated washing procedure all apparatus must be rinsed in deionised water before drying. Deionisation may be achieved by passing the water through a cartridge available from Elgastat Ltd., or Millipore Corp. (Appendix 3). It is important to maintain the cartridge in good condition and its effectiveness should be checked daily.

4.1.1. Washing procedure

A number of different procedures are in use, but it is essential to avoid the use of abrasive and highly caustic cleaning agents such as *Vim* or concentrated acids. The alkaline sodium metasilicate is a very suitable cleaning agent and has the advantage that if a layer of metasilicate remains it is deposited as glass on neutralisation. *Calgon metasilicate* (CMS) is made up as follows. Dissolve 360 g sodium metasilicate (British Drug Houses, technical grade) and 40 g calgon (a water softener, available from R. and J. Wood, Ltd.) in 1 gallon of hot water. Dilute 1 to 100 with water for use.

As an alternative to CMS the detergents *Decon* or *7X* may be used. These come as concentrated liquids and are more convenient to use than CMS. Decon 90 (phosphate free) must be used in hard water areas but Decon 75 is suitable in soft water areas. Decon (available from Decon Labs Ltd.; Appendix 3) should be used as a 2% solution. 7X (available from Flow Laboratories; Appendix 3) should be used as a 5% solution. They are anionic surfactants.

4.1.2. Bottles and pipettes

1. Glassware is removed from the soaking baths and cotton wool plugs removed from pipettes (take care not to splash the caustic chloros).
2. Place calgon metasilicate (CMS) in a boiler and boil for 20 min. Alternatively bottles may be filled with hot CMS, Decon or 7X and left overnight.
3. Allow to cool and rinse *six* times in tap water.

4. Rinse *three* times in deionised water.
5. Dry in a hot air oven – if any white streaks are found on the sides of the vessels the rinsing procedure is inadequate. This may be a result of failure to rinse sufficiently or it may indicate that the deionised water supply is faulty.

4.1.3. Rubber stoppers

Natural or silicone rubber stoppers and capliners should be boiled in CMS, Decon or 7X, rinsed well in tap water and boiled in deionised water for 10 min. After further rinsing in deionised water they should be dried at low temperature. The metal caps should not be treated with CMS but may instead be boiled in a 1% bicarbonate solution.

4.1.4. Glass washing machines

If a suitable glass washing machine is available then after stage 2 described above, the glassware may be loaded into the machine for a further wash and rinsed using deionised water. Addition of detergent at this stage is usually unnecessary but if required powdered Decon or 7X-omatic may be used. 7X-omatic (Flow Laboratories; Appendix 3) is a low foaming liquid containing a non-ionic surfactant and should be used as a 1% solution.

4.2.   Sterilisation methods

The following are methods of sterilisation:
   Hot air
   Autoclaving
   Filtration
   Irradiation e.g. Co 60 γ-irradiation is used commercially for disposable plastics
   Ethylene oxide treatment
The last two methods are not generally applicable to the small laboratory except insofar as low pressure mercury vapour lamps emitting light of 254 nm may be used to sterilise the air in aseptic rooms and cabinets (see § 9.4.1).

4.2.1. Hot air

This involves heating the glassware to 160°C for 90–120 min. It is the preferred method for bottles which do not have screw caps when their orifice should be covered with aluminium foil. It is also used for glass Petri dishes and pipettes. Petri dishes should either be stacked in tins or containers specially designed for them or simply held closed with *sterilisation tape*. This brown paper adhesive tape has pale stripes which turn dark brown under sterilising conditions and is used as an indicator of the effectiveness of the procedure. It is available from the 3 M Company Ltd. (Appendix 3).

Pipettes are plugged with non-absorbent cotton wool and placed in aluminium or steel cans (obtainable from Gallenkamp or A. and J. Beveridge; Appendix 3) for sterilising. At the bottom of the can a pad of cotton wool may be placed to prevent damage to the pipette tips. This pad should be renewed each time the pipette can is sterilised as prolonged heat causes the cotton wool to carbonise producing material toxic to cells. A Browne's steriliser tube Type III (see below) should be placed inside the central can of each batch being sterilised.

4.2.2. Autoclaving

Heat-stable solutions, rubber bungs and liners, bottles with plastic caps, ultrafiltration apparatus etc. are all sterilised by steam treatment at elevated pressure. Although the time required to sterilise is usually only about 15 min at 15 lb pressure the cycle time for modern autoclaves is several hours. This is because of the safety precautions built into these machines to prevent the doors being opened until the temperature of liquid within bottles has fallen to 80°C.

To sterilise small items such as rubber bungs they should be placed in glass Petri dishes and wrapped in aluminium foil.

Empty bottles should have their plastic caps only loosely screwed on to allow penetration and escape of steam during the sterilisation cycle.

Partly filled bottles should have their caps firmly screwed on.

The steam generated within the bottle will effectively sterilise the contents but changes in volume will be prevented. It is such bottles which create the hazard in autoclaves if the temperature and pressure outside the bottle is allowed to fall suddenly. This will happen if the autoclave door is opened prematurely leaving the bottle itself full of superheated steam.

To obtain sterile gauze for filtering trypsinised suspensions when setting up primary cells the gauze is best arranged in a filter funnel and the whole wrapped in aluminium foil so that it can be unwrapped without contamination. Such wrapped funnels are sterilised by autoclaving.

Filtration apparatus (see below) can be assembled with the filters in position and the whole autoclaved. It is important either to plug orifices with cotton wool or to cover them with aluminium foil caps.

When it is desired to sterilise only a small amount of material the use of an autoclave may be extravagant. In this case an ordinary pressure cooker with facilities to go up to 15 lb pressure may be used, but precautions must be taken to allow adequate time for cooling before the cooker is opened.

4.2.3. Control of sterilisation

Browne's steriliser tubes contain a liquid that undergoes a slow chemical reaction at elevated temperatures, changing from red to amber then green. Only when they are green has the correct temperature time combination elapsed. Several of these indicator tubes should be distributed in each batch of material to be sterilised. It is important when autoclaving liquids, to place a Browne's tube in a control bottle of liquid as the conditions within such a bottle differ from those outside. Different types of Browne's tubes are available for sterilisation using steam or dry heat. Type I (black spot) tubes are for use for steam sterilisation and type III (green spot) for sterilisation by dry heat. They are available from Albert Browne Ltd. (Appendix 3) and each box comes with instructions for use and a colour code.

4.2.4. Filtration

Membrane filters remove, from a liquid passing through them, all particulate matter larger than the filter pores. They are of high porosity giving reasonably high flow rates and by selection of the appropriate pore size the resulting filtrate is rendered sterile. It is usual to have a 0.22 μm pore, though in some cases smaller pores are required and often two or even three membranes of decreasing pore size are used, e.g. 0.45, 0.22 and 0.10 μm. The filters are made of cellulose esters and so may be sterilised by autoclaving. They are first fitted into a filter holder and where necessary covered in aluminium foil. They are available in a large range of diameters and so filtration is a convenient method of sterilising volumes of liquid ranging from less than 1 ml to 5 gallons or more.

For small volumes, 13 or 25 mm diameter, filtration membranes may be fitted into plastic or stainless steel holders (e.g. the Swinnex filter holder made by the Millipore Corp.; Appendix 3) which, after autoclaving, are fitted onto a syringe containing the liquid to be sterilised. Care must be taken in the assembling of the membrane in the holder as incorrect assembly leads to the escape of the membrane from its retaining gaskets with subsequent failure of the filtration process. The plastic holders have a limited life time as they distort on autoclaving. A more reliable, though more expensive method is to buy filters presealed into sterile disposable holders (Millex filter units available from Millipore Corp.; Appendix 3).

For volumes up to about 300 ml stainless steel filter holders are available (see below). These take a 47 or 50 mm diameter membrane filter and incorporate an upper chamber which holds either 100 ml or 330 ml of the liquid to be sterilised. Once assembled the unit is sterilised by autoclaving with the lower outlet covered with aluminium foil. In use the unit is clamped in a vertical position with the outlet over a sterile receiving vessel. The liquid to be sterilised is placed in the upper compartment and pressure is applied to force the liquid through the membrane. The pressure may be applied from a cylinder of compressed air or by means of a small air pump. Using a 0.22 μm membrane filter and a pressure of 10 psi a flow rate of 140 ml/min can be achieved. To reduce the chance of contamination a neoprene stopper may be fitted to the outlet tube and this fitted into a Buchner flask with a cotton wool filter fitted to the side arm. The whole assembly may be autoclaved together. Another method is to fit a 'bell' to the outlet tube (Fig. 4.1).

The volume of liquid that may be filtered through such a filter can be increased by coupling the holder to a dispensing pressure vessel. Vessels holding up to 20 l may be interposed between the smaller filter holder and the supply of compressed air so that liquid is constantly transferred from the large reservoir to the small reservoir above the membrane. However, a 47 mm diameter filter often clogs up before filtration of such large volumes is complete and it is recommended that a large, e.g. 142 mm diameter, filter is used. 142 mm diameter filter holders do not have a built-in reservoir and must be coupled to a dispensing pressure vessel. However, flow rates of up to 1.5 l/min can be obtained with a pressure of 10 psi and the holder comes fitted with legs which conveniently can sit astride the collecting vessel.

These stainless steel filter holders are expensive and require practice in order to assemble them correctly. They also require careful washing and sterilising, and a new type of disposable unit is now available from Millipore Corp. (Appendix 3). The 'twin 90' filter unit has the same filter area as the 142 mm filters achieved by scaling two 90 mm, 0.22 µm Millipore filter discs in parallel into a clear plastic housing. The unit is pre-sterilised and comes with a detachable filling bell. All that is required is to attach it to a dispensing pressure vessel. Although very expensive, the cost of these units must be viewed in terms of their labour saving potential.

If it is necessary to refill the dispensing pressure vessel during filtration it is important that the pressure above the membrane filter does not fall or contamination may result. To avoid such a pressure fall valves may be fitted. Foot operated systems are available from Schleicher and Schuell which allow the flow of liquid to be interrupted at regular intervals; this is very useful when filling 100 ml

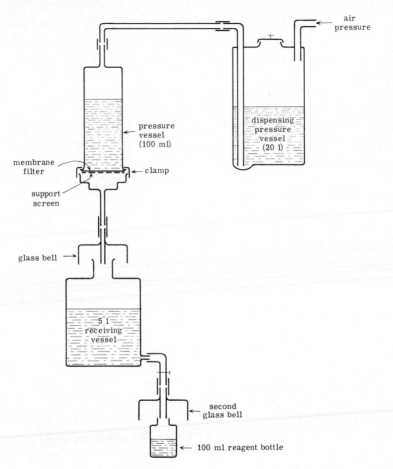

Fig. 4.1. Diagrammatic outline of the arrangement of vessels used in filter sterilisation. When filters of larger diameter are used the filter holder is not designed to hold any liquid and a dispensing pressure vessel must be used. However, with the 100 ml filter holders the dispensing pressure vessel is unnecessary if only small volumes are to be filtered. In this case the filtrate may be collected directly into a sterile reagent bottle.

bottles from an otherwise continuous flow of 10 l of sterile medium. However, we have found the system cumbersome to use and we prefer to collect the sterile medium first into a large, sterile bottle and subsequently withdraw 100 ml aliquots using a rubber hose and a second filling bell (see Fig. 4.1).

Although the Millipore Corporation offer a very wide range of filter holders and accessories, other companies provide valuable competition: e.g. Anderman and Co., Ltd. (Appendix 3) are agents for Schleicher and Schnell who manufacture a comparable range of filters and holders etc., and V.A. Howe supply Sartorius membranes and filter holders.

A 0.22 μm sterile disposable filter unit with a capacity of 150 ml or greater is available from Falcon (Becton-Dickinson, U.K. Ltd.; Appendix 3). This unit resembles a sterile beaker to the top of which is sealed the filter. A reservoir sits above the filter. A vacuum line is attached to an outlet on the side of the receiving beaker so that filtration is under negative pressure. Because of this it is not very suitable for filtration of proteinaceous solutions. It is a very convenient method, however, of filtering other solutions, and the filtrate is obtained in a portable form which is not readily contaminated.

CHAPTER 5

Subculturing

5.1. Dissociation techniques

Cells in tissues and cells growing as monolayers on glass or plastic surfaces are held together and to the substratum by mucoproteins, and sometimes by collagen (see § 2.3.2). In addition, many cell monolayers and tissues, especially epithelial tissues, require divalent cations (Ca^{2+}, Mg^{2+}) for their integrity.

Thus for tissue dissociation or for releasing cells from monolayers various protease solutions are used, sometimes in association with chelating agents.

5.1.1. Trypsin

For many years a solution of a crude acetone powder of bovine or porcine pancreas has been used to disaggregate tissues and to release cultured cells from their substratum. These preparations – referred to as trypsin 1:250 (based on an international standard) – contain not only trypsin but a range of enzymes including chymotrypsin and elastase which are equally important. Purified trypsin seldom is as efficient as the cruder preparations, especially on tissue disaggregation, when mucinous clumps result (Ronaldoni, 1959).

Furthermore, prolonged exposure to trypsin should be avoided as this damages the cells. The best way to inactive the trypsin when the desired dissociation has been achieved is to add serum (or medium containing serum) which contains a natural trypsin inhibitor.

Trypsin solutions are usually made up in a saline solution (PBS-A) (see Appendix 1) or in saline–citrate and are used as 0.25% solutions.

To make 1 l trypsin–citrate dissolve trypsin (Difco 1:250) as follows by vigorous stirring for 3–4 h:

trypsin (1:250)	2.5 g
tri-sodium citrate	2.96 g
sodium chloride	6.15 g
phenol red (1%)	1.5 ml
distilled water	to 1 l

Adjust the pH to 7.8 with NaOH (approx. 2 ml 1 N NaOH) and filter several times through Whatman's No. 1 filter paper. Sterilise by filtration using a prefilter and an 0.22 μm pore size membrane.

This stock solution is conveniently stored frozen at $-20°$ in 20 ml and 80 ml amounts after checking for contamination using 1) Saboraud fluid medium at 31°C for 1 week (Appendix 4); 2) brain heart infusion broth at 37°C for 1 week (Appendix 4). (One check of each should contain calf serum.)

5.1.2. Pronase

Pronase is recommended for primary cultures as it gives a better single cell suspension than trypsin (Gwatkin 1973). It is not always as good with cell lines. An 0.25% solution is made by dissolving 1.25 g standard B grade pronase (Calbiochem Corp.) in 500 ml PBS-A at room temperature. After 30 min the cloudy solution is centrifuged (100 g, 10 min) and sterilised by filtration as for trypsin.

5.1.3. Collagenase

Collagenase is reported to cause least damage and is used in the preparation of clonal cell cultures (Hilfer, 1973).

Combined use of collagenase and hyaluronidase followed by trypsin or pronase without vigorous agitation (which leads to loss of cell viability) was satisfactory for mouse mammary gland cells (Prop and Wiepjes, 1973). Preparations of collagenase vary. Worthington (Appendix 3) supply four types of crude collagenase selected for their suitability in preparing cells from different tissues, and Sigma Chemical Co. (Appendix 3) provide two different types for isolation of cells from liver or fat pads.

5.1.4. Versene (EDTA)

Treatment with chelating agents such as EDTA which removes divalent ions, leads to dissociation of cell monolayers and release of the cells into suspension without protease action. Perfusion of rat liver with citrate solutions has also been used in attempts to produce primary liver cell cultures.

Very often, however, mixtures of trypsin or trypsin-citrate with versene are used to release cells from monolayer cultures as the combination works very much faster than either alone. In our laboratory the following procedure is used to make up 10 l of versene in PBS-A:

	10 litres
NaCl	80 g
KCl	2.0 g
Na_2HPO_4	11.5 g
KH_2PO_4	2.0 g
Versene (diaminoethanetetra-acetic acid disodium salt)	2.0 g
Phenol Red 1%	15 ml
Distilled water to	10 l

Dispense in 20 ml, 80 ml and 160 ml amounts.
Autoclave at 15 lb pressure for 15 min for 20 ml bottles.
Autoclave at 15 lb pressure for 20 min for 80 ml and 160 ml bottles.
Store at room temperature.

To make trypsin–versene mix:
  1 volume 0.25% trypsin or trypsin–citrate (see above)
  4 volumes versene

5.1.5. Mechanical means

For many biochemical studies it is undesirable to release the cells from the substratum using trypsin. This is especially true in two circumstances.

  1. Studies on cellular surfaces. Trypsin leads to loss of surface proteins including glycoproteins and other antigens and it is these

changes which presumably lead to cell death when trypsinisation is prolonged.

2. Studies involving timed incubations with drugs, radioactive precursors, hormones, etc. In these instances it is important to harvest the cells promptly without exposure at 37°C to an altered environment for an indefinite period.

For these studies cells must be harvested mechanically, e g by scraping with a rubber policeman or by the use of chemicals (e.g. alkali, acid or detergent) which lead to instantaneous death of the cells from which various products may then be isolated.

Rubber policemen are simply rubber sleeves (obtained from Mac-Farlane Robson Ltd.; Appendix 3), which fit over the ends of glass rods and provide a soft surface with which cells may be scraped from their substratum. Alternatives for use with dishes are wedges of silicone rubber cut from bungs and stuck on hypodermic needles (Fig. 5.1). These also have the advantage that they can be readily sterilised by autoclaving. A collapsible type of windscreen wiper (Fig. 5.1) is readily constructed for scraping cells from the inside of roller bottles.

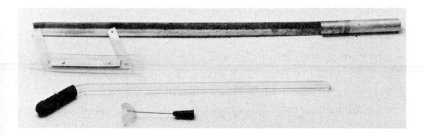

Fig. 5.1. Scrapers used to remove cells from surfaces. The upper scraper is used to scrape cells from inside roller bottles. The blade folds forwards allowing it to pass through the narrow neck. By pushing against the bottom of the bottle the blade opens as shown and the shape of the arm prevents further movement. The middle scraper is a rubber policeman fitted to a bent glass rod. It is used for scraping cells from smaller bottles. The lower scraper is used for dishes where the edges of the cut silicone bung allow cells to be removed efficiently from the corners of the dish.

5.2.   Subculture of a cell monolayer

a. Prepare a bottle of suitable growth medium and warm to 37°C.
b. To 20 ml versene add 5 ml trypsin–citrate solution and warm to 37°C.
c. Examine the monolayer macroscopically and microscopically for any signs of bacterial contamination (Chapter 9).
d. Aseptically pour off the medium into a universal bottle. This is for a bacteria check (see below).
e. Add a small volume (1 ml for a 50 ml bottle; 5 ml for a Roux bottle) of trypsin–versene solution at 37°C.
f. Immediately remove this trypsin–versene and add a further small volume of fresh trypsin–versene at 37°C.
g. Replace the cap and incubate the bottle at 37°C. After a few minutes the cells become released into suspension. This process may be aided by *gently* tapping the base of the bottle with the palm of the hand.
h. DO NOT LEAVE THE CELLS IN TRYPSIN FOR LONGER THAN NECESSARY AS IT WILL DAMAGE THE CELLS.
i. To prevent excessive trypsin action add 5–20 ml growth medium containing serum.
j. Pipette gently to complete dispersal and count the cells (see below and § 8.2).
k. Transfer the appropriate volume of cell suspension (see Table 3.1) into new bottles and dishes and add fresh medium. Usually cells may be split 1:4 every 2–3 days but some strains grow better if split 1:2 and some cell lines may be split 1:10.
l. Gas the bottles with 5% CO_2 in air (see § 3.1.2) and incubate at 37°C.

5.2.1. Viable cell count

Take some cell suspension and dilute with an equal volume of 0.1% trypan blue (stains dead cells: Appendix 2).

Take up some trypan blue cell suspension in a Pasteur pipette and fill a haemocytometer counting chamber by capillary attraction

(§ 8.2.2). Take care not to flood the channels of the chamber. Count the cells under × 10 objective.

Count the four sets of sixteen small squares (on each corner of the central ruled area (§ 8.2.1). Count only unstained cells. Divide by 4 to give the average count per 1 mm^2. Since the area is 1 mm^2 and the depth 0.1 mm, the conversion factor for the counting chamber is 10^4, e.g. let the average count = 20 cells per 1 mm^2

$$20 \times 10^4 = \text{number of cells per ml of trypan blue suspension}$$

$$2 \times 20 \times 10^4 = \text{number of cells per ml of original suspension}$$

i.e. 4×10^5 cells per ml.

5.2.2. Bacterial check

Take the medium from step d of the subculture procedure and centrifuge at 800 g for 15 min. Discard the supernatant. Using a platinum loop, place the sediment on a blood agar plate (Appendix 4) and incubate at 37°C for at least 2 days to check for bacterial growth (Chapter 9).

5.3. The growth cycle

On subculturing cells into new vessels they do not start to grow immediately at their maximum rate especially if they are seeded as a dilute inoculum. They rather exhibit a lag phase of 1–2 days while they adapt to their new environment and condition the new medium. That this is not an intrinsic characteristic of cells was shown by Puck (1972). The lag phase can be almost entirely eliminated by taking care not to over-trypsinise and to maintain the cells at 37°C, and with mouse L929 cells most cells will divide within a day when more concentrated inocula are used. However, it is often found that little increase in cell number occurs on the first day following subculture; only after two days have the cells doubled in number. Thereafter, exponential growth may be maintained for a further 2–10 days if there is sufficient room in the vessel and the medium is changed regularly. Before long, however, the rate of growth falls and the cell number plateaus (Fig. 5.2). Some cells remain viable in this state for up to several weeks, especially if the medium is changed every few days, but with other cell lines growing in so-called monolayer culture the cells pile up one on top of the other, the lower cells become starved and soon the cell sheet peels off the glass. Changing the medium in stationary monolayer cultures prevents the build-up of acid and other waste products and supplies fresh nutrients. In cells which have just entered the plateau it also has the effect of stimulating some of the cells to undergo a new

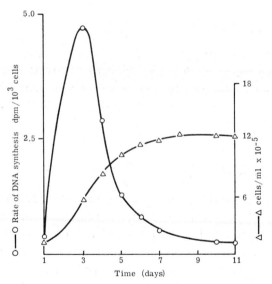

Fig. 5.2. Growth cycle of mouse L929 cells. Mouse L929 cells were inoculated into 2 oz medical flat bottles (5×10^5 cells in 5 ml Eagle's Minimal Essential Medium containing 10% calf serum). The medium was changed every two days. Bottles were incubated for 60 min with 2 μCi (6-^{3}H)-thymidine (80 μCi/μmol) at a final concentration of 5 μM after which they were harvested by trypsinisation, their number (△———△) estimated using a Coulter counter and the rate of DNA synthesis (O———O) estimated from the incorporation of [^{3}H]thymidine (Chapter 12). (Courtesy of Dr. J.G. Lindsay, 1969.)

round of DNA synthesis and cell division. This is particularly marked in mouse L929 cells as regular feeding enables a very high density monolayer to form in which all the cells are of a small uniform size.

With monolayer cultures it is uncommon to extend the length of the experimental phase of growth for more than a few days. Seeding the cells at lower densities usually leads to a longer lag phase during which some cell division occurs but the generation time is extended and in general an eightfold increase in cell numbers (i.e. 3 cell divisions) is all that is achieved during the exponential phase. However, as stated above, Puck (1972) has achieved reproducible exponential growth from single HeLa cells which continued for more than 10 days and gave rise to a 2000-fold increase in cell number.

In suspension, however, cells may be maintained in exponential growth for long periods, and in theory indefinitely if the cells are grown in a chemostat where the small trickle of incoming nutrients is balanced by the outflow of cell suspension thus maintaining a constant environment and constant cell number. In practice this is difficult to achieve, and contamination is a serious hazard. Cells in suspension are usually grown in batch cultures which exhibit similar growth kinetics to monolayer cultures.

CHAPTER 6

Primary cells

6.1. Introduction

Cells taken from an animal and placed in culture are termed primary cells until they are subcultured. This chapter is concerned with the methods used to obtain cells from animals. As far as humans are concerned it is important to obtain cells while bringing about minimum discomfort to the donor. For this purpose two cell types are commonly used: they are the lymphocyte and the skin fibroblasts. The lymphocyte is in general a non-dividing cell and requires mitogenic stimulation, usually by phytohaemagglutinin or other plant lectin, when it undergoes at most a few cell divisions before dying. The skin fibroblast also has a limited, but far longer, lifespan (Hayflick, 1965a) but is seldom available in the same quantity as the lymphocyte.

6.2. Lymphocytes

Lymphocytes have been isolated from a variety of mammalian sources. The horse is an ideal source if a large (200 ml) regular supply of lymphocytes from one donor is required (Zain et al., 1973) but human lymphocytes are an ideal source of human tissue for the study of genetic diseases (Abo-Darub et al., 1978). The use of pig lymphocytes (Forsdyke, 1968) usually involves slaughter of the animal.

6.2.1. Isolation of leukocytes and autologous plasma

Blood is taken aseptically into a medical flat bottle containing heparin (2500 I.U./100 ml blood). 'Pularin' (Evans Medical Co.; Appendix 3) is a suitable source of heparin without preservative. The blood is allowed to stand at room temperature for 40 min and the leukocyte-rich plasma is then withdrawn aseptically and centrifuged at 400 g for 15 min at room temperature. The supernatant is withdrawn and recentrifuged at 1000 g for 10 min at room temperature to give a cell free autologous plasma preparation which can be stored at $-20°C$. The cell sediment is resuspended in 1/10th the initial volume of autologous plasma.

6.2.2. Purification of lymphocytes

Two different methods are in general use: the glass bead column method of Rabinowitz (1964, 1973) and the density gradient centrifugation method of Boyum (1968).

6.2.3. Glass-bead column method

A sterile glass condenser or water jacketed column (40 × 2 cm) is clamped in a vertical position and water at 37°C is circulated around it. This size of column is large enough to fractionate leukocytes from 200 ml of blood.

Glass beads (Type 070-5005 superbrite brand from 3M Company, St. Paul, Minnesota) are soaked overnight in concentrated HNO_3 and then rinsed exhaustively in tap and distilled water before drying and sterilising at 160°C for 3 h. The sterile beads are poured into the column to fill it 3/4 full and then washed with Eagle's MEM (pregassed with 5% CO_2).

The leukocyte suspension is applied to the top of the column and allowed to percolate in. Considerable losses may occur at this stage as the cells stick to the tubes and pipettes, but this can be countered by using siliconised glassware (e.g. treated with Repelcote; Appendix 3). The cells are left on the column at 37°C for 30 min and then eluted with Eagle's MEM supplemented with 50% autologous plasma.

Most of the lymphocytes and any erythrocytes are eluted in the first column volume, closely followed by the platelets and usually the first 40 ml is collected from a 40 × 2 cm column. Granulocytes and monocytes are only eluted with 0.02% EDTA in PBS-A. Failure of this method may be associated with the use of aged serum or plasma and hence the fresh autologous plasma is recommended. If only limited amounts of this are available its concentration may be reduced. The contaminating red blood cells may be removed by hypotonic lysis but this sometimes leads to destruction of some lymphocytes (Dain and Hall, 1967; Roos and Loos, 1970).

6.2.4. Gradient centrifugation method

Triosil (Na-metrizoate or Na-N-methyl-3,5-diacetamido-2,4,4-tri-iodobenzoate) is provided as a radiographic contrast medium (60% (w/v) solution containing 55.2% (w/v) Na-metrizoate, 2.8% (w/v) Ca-metrizoate and 2% (w/v) Mg-metrizoate) by Glaxo Laboratories, Ltd., (Appendix 3). The density of this solution is 1.39 g/ml and it is diluted with water to 1.2 g/ml. Just before use 10 parts are further diluted with 24 parts 9% (w/v) Ficoll (Pharmacia Chemicals, Uppsala; aqueous solution sterilised by autoclaving) to give 'separation fluid'. A readily prepared mixture 'Ficoll–Paque' is available from Pharmacia Chemicals (Appendix 3).

10 ml of leukocyte suspension is diluted with 20 ml sterile 0.9% NaCl and 28 ml layered over 10.5 ml of separation fluid in a sterile 14 × 3 cm tube. Smaller quantities may be used but the height of the layers should be approximately the same. After centrifugation at 400 g for 25 min at 20°C the red cells are clumped by Ficoll and sediment to the bottom of the tube while the lymphocytes form a white band at the junction of the separation fluid and saline layers. This is removed with a syringe and the cells are washed twice by resuspension in Eagle's MEM (Appendix 1) supplemented with serine (10 μg/ml) and glycine (7.5 μg/ml), and centrifugation at 800 g for 10 min. The cells are finally resuspended in this modified Eagle's MEM containing 10% autologous plasma. A differential white blood

cell count (Hunter and Bomford, 1968) shows that 80–90% of the cells are small lymphocytes.

The standard method recommended by Pharmacia (Appendix 3) uses 2 ml whole anticoagulant-treated blood diluted with an equal volume of BSS. This is layered over 3 ml 'Ficoll–Paque' and sedimented at 400 g for 30 min. The lymphocytes may be removed from the interphase and the upper layer may be used as a source of autologous plasma.

An alternative dense medium for fractionating blood cells is 'Percoll' also provided by Pharmacia (Appendix 3). Percoll consists of particles of colloidal silica coated with polyvinylpyrrolidone (Pertoft et al., 1978). It is made up by dilution into isotonic saline and forms a self-generating gradient on centrifugation at 20,000 g for 15–20 min in an angle rotor. Onto this gradient may be layered a sample of anticoagulant-treated blood which can be fractionated in two stages. The first stage involves centrifuging for 5 min at 400 g and removing the plasma layer containing the platelets from the top of the gradient. The second stage involves centrifuging for 15 min at 800 g when lymphocytes, granulocytes and erythrocytes band at their isopycnic densities. They can be harvested by upward displacement using 60% sucrose and diluted and washed with saline solutions by sedimentation at 400 g.

6.2.5. Cultured lymphocytes

The cells, suspended at 10^6 per ml of Eagle's MEM supplemented with serine and glycine and 10% autologous serum and containing 50 I.U. of penicillin and streptomycin per ml, are added to 12 × 1.3 cm capped tubes (1 ml), 6 cm Petri dishes (2 ml), 14 × 3 cm tubes (5 ml) or roller bottles (50 ml upwards). The vessels are either loosely capped and incubated in a CO_2 incubator or are gassed with 5% CO_2 in air and sealed prior to incubation.

To induce blastogenesis an equal volume of medium containing mitogen is added. The most commonly used mitogen is phytohaemagglutinin (PHA). 100 mg PHA-M (Difco Laboratories; Appendix 3) is rehydrated with 5 ml sterile distilled water and further diluted 1 in 200 with growth medium. Pokeweed mitogen (Gibco Biocult, Ltd.) is reconstituted in a similar manner but only diluted 1 in 40 with growth medium to give an active solution.

Immediately on adding mitogen the cells begin to clump and it is for this reason that it is preferable first to dispense the cells into their individual vessels as reproducible aliquots are not easily obtained once the cells have clumped. Under these conditions DNA synthesis – as measured by incorporation of radioactive thymidine into acid- and ethanol-insoluble material (see § 12.2) – starts around 24 h after mitogen addition and increases in rate over the next 2 or 3 days. This labelling of DNA may be for 6 h periods with [^{14}C]thymidine (2 μCi/ml [2-^{14}C]thymidine, 3.7 Ci/mol) or for 2 h periods with

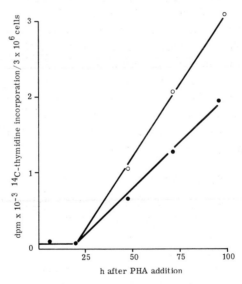

Fig. 6.1. DNA synthesis in PHA-stimulated horse lymphocytes. Aliquots of 3×10^6 horse lymphocytes purified by the glass bead column method were incubated for various lengths of time with PHA-M in tubes in 3 ml Eagle's MEM (supplemented with 10 μg serine/ml and 7.5 μg glycine/ml and 10% autologous horse plasma). Before harvesting the cells were labelled for 6 h with [^{14}C]thymidine (3.66 Ci/mol; 1 μCi/ml). O, Amethopterin (10^{-6} M) was added 16 h before harvesting; ●, no amethopterin added. (Courtesy of Dr. B.S. Zain, 1971.)

[^{3}H]thymidine ([6-^{3}H]thymidine at 4 μM) after which the cells are sedimented, washed three times with 5% trichloroacetic acid, twice with absolute ethanol, dried with ether and solubilised in hyamine hydroxide for counting in a toluene/PPO scintillator (see § 12.2.2.2). The sensitivity of the assay may be increased by causing partial synchrony of DNA synthesis. This may be achieved by pretreatment of the cells with Amethopterin (5 × 10^{-7} M) for several hours before and during the exposure to radioactive thymidine (Tormey and Mueller, 1965; Pegoraro and Benzio, 1971) (see Fig. 6.1).

6.2.6. Separation of T and B lymphocytes

A method described by Hellström et al. (1976) makes use of the selective adsorption of neuraminidase-treated human T lymphocytes to Helix pomatia lectin conjugated to Sepharose 6 MB (extra large bead size). The lymphocytes purified by gradient centrifugation (§ 6.2.4) are treated with neuraminidase from *Clostridium perfringens* (5 μg/ml, 45 min, 37°C) and then washed in PBS containing human serum albumin (HSA 0.2%) and sodium azide (NaN$_3$ 0.02%).

About 10^8 treated lymphocytes are allowed to permeate 3 ml of lectin–Sepharose in a small column. After 15 min, unbound cells (largely B cells as shown by the presence of surface immunoglobulin and receptors for complement-activated sheep erythrocytes) are eluted with 80 ml PBS–HSA–NaN$_3$ (8 ml/min). A mixed population of weakly bound cells is removed with a further 80 ml of PBS–HSA–NaN$_3$ containing 0.1 mg/ml N-acetyl-α-D-galactosamine and a preparation greatly enriched in T lymphocytes is finally eluted with PBS–HSA–NaN$_3$ containing 1.0 mg/ml N-acetyl-α-D-galactosamine.

Sodium azide is removed from the cells by washing prior to use.

6.3. Human skin biopsies

– Select an area of skin – usually on the inner side of the upper arm – which is free of hair and minimally keratinised.
– Sterilise by swabbing with 70% ethanol and pull a piece of skin up with forceps.

- With sharp scissors excise 1 mm³ of the skin and place in a 6 cm Petri dish.
- Add a drop of complete growth medium (e.g. Dulbecco's MEM containing 10% foetal calf serum) and cut the skin fragment into 5–10 pieces.
- Dispose the several pieces over the surface of the dish and cover each with a drop of complete medium.
- Incubate overnight at 37°C. Sometimes this incubation is done with the dish tilted so that the medium rests at the edge of the dish, thereby maintaining high humidity yet encouraging attachment of the tissue fragments to the surface of the dish.
- Very carefully, so as not to dislodge the fragments, add 3 ml complete medium to the dish and continue the incubation.

Alternatively, rather than using dishes, the tissue fragments may be transferred to a small bottle and 5 ml complete medium may be

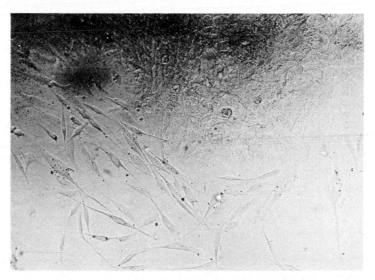

Fig. 6.2. Primary cells growing out of human skin explant. Seven days after fragments of human skin are placed in culture fibroblasts and epithelial cells can be seen growing out of the explant. (Courtesy of M.L. Hodgins.)

added in such a way that the bottle can be incubated with the fragments adhering to the underside of the upper surface. When the fragments have attached after 16 h the bottle is turned carefully so that the medium now bathes the fragments.

Fibroblasts grow out from the tissue fragments within a week (Fig. 6.2).

6.4. Primary kidney cells

This method is applicable to kidneys of various mammals when allowance is made for scale.

- Remove the kidneys aseptically. In order to do this it will be necessary to shave the animal and then swab the skin with 70% ethanol before making the incision. The dissecting instruments should be dipped into ethanol and flamed before use.
- Rinse the kidneys in sterile BSS and transfer to a sterile Petri dish for dissection.
- Remove the capsule. This may be done by holding the kidney on edge and slicing it into two halves with a scalpel. The capsule is not severed at the bottom and the two kidney halves may be peeled away from it.
- Using fine scissors separate the cortex from the medulla and discard the latter.
- Chop up the cortex into a mush using a razor blade held in artery forceps.
- Flood the chopped tissue in BSS (Appendix 1) and remove the BSS with a Pasteur pipette.
- Transfer the tissue to a flask with an equal volume of 0.25% trypsin in PBS-A at 37°C (Appendix 1). The flask may be a special trypsinisation flask obtainable from Bellco Glass Inc., NJ, or may be a Virtis or MSE homogenisation flask.
- Stir mechanically at 37°C at maximum speed possible without causing frothing.
- After 0.5–1 h allow the tissue to settle and decant the cell suspension. This should be stored on ice until the next step.

- Further trypsinisation cycles (2 or 3) are carried out until all the tissue is disaggregated.
- Sediment the cells from the pooled suspension by centrifugation at room temperature at very low speed (50–100 g for 20–30 min).
- Remove the supernatant fluid using a sterile Pasteur pipette attached through an aspirator bottle to a water pump.
- Resuspend the cells in prewarmed growth medium and filter through muslin or cheesecloth (filter funnels with muslin held in place with autoclave tape can be wrapped in aluminium foil or Kraft paper and sterilised by autoclaving).
- A yield of about 10^8 cells/g of cortical tissue can be obtained and these should be diluted to 3×10^5 cells/ml and distributed into

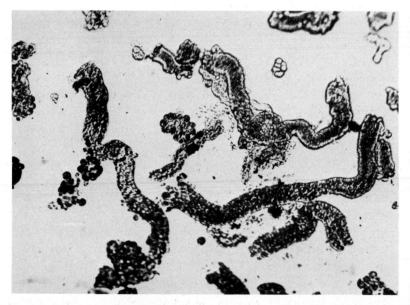

Fig. 6.3. Rabbit kidney tubule fragments. In the medium devised by Lieberman and Ove (1962) these tubule fragments immediately attach to the glass surface and growth is induced so that within 2 days a monolayer of cells is formed. (Reproduced from Lieberman and Ove, 1962, with kind permission of the authors and publishers.)

flasks for growth. Melnick's medium A (Melnick, 1955) was designed for growth of primary monkey kidney cells (see Chapter 7).

A variant of this procedure was devised by Lieberman and Ove (1962). This led to the isolation of rabbit kidney tubule fragments rather than individual cells (Fig. 6.3) and a special serum-less medium was devised for their cultivation. Within two days the cells grew out from the tubule fragments to form a cell monolayer. This system proved very useful in investigating the biochemical changes accompanying the changeover from a non-growing population of kidney cortex cells in vivo to a dividing in vitro population (Lieberman et al., 1963; Adams et al., 1966; Lee et al., 1970).

6.5. *Mouse macrophage cultures* (see Nelson, 1976)

– Kill the mouse (6-week-old mice are best) by cervical dislocation and pin to a board.
– Spray with 70% ethanol and remove the skin from the abdominal wall. Do not puncture the abdominal wall.
– Inject 2.5 ml growth medium (Eagle's MEM plus 10% calf serum and heparin at 10 U/ml), together with some air into the peritoneum. Do not puncture the gut. The abdomen balloons up sealing the site of injection.
– Massage the abdominal wall for 1 min to release macrophages into suspension.
– Using a syringe fitted with a 21 g × 1½″ needle aspirate the fluid. Insert the needle-cyclet inwards, into the flank and pull sideways to prevent blocking.

Count the cells and distribute into dishes (10^6 cells/5 cm dish). Macrophages will attach within 30 min and contaminating lymphocytes and fibroblasts may be removed. Peritoneal macrophages do not normally grow in vitro unless conditioned medium (§ 8.1.1) is used. Mauel and Defendi (1971a) have isolated a macrophage growth factor from conditioned medium and have also obtained some permanent macrophage cell lines (Mauel and Defendi, 1971b).

6.6. *Rat or chick skeletal muscle cells*

(1–2-day-old rats must be used as older rats yield only fibroblasts.)
Remove the thigh muscle and trypsinise in a manner similar to rabbit or monkey kidney. The final cell suspension should be filtered through a double layer of sterile lens cleaning tissue fixed in a Swinnex Millipore filter holder (see § 4.2.4) mounted on a 20 ml syringe (Yaffe, 1973). For differentiation the culture vessel must be coated with collagen or gelatin: dissolve gelatin to 0.01% in hot distilled water and autoclave. Put 3 ml into a 6 cm dish and leave in the cold for 2 h. Remove the gelatin solution and add the cell suspension in a mixture of medium 199 (Morgan et al., 1950) and Dulbecco's Modified Eagle's medium (1:4) with 10% horse serum and 1% chick embryo extract (Appendix 1).

Cells plated at 2–3.5 × 10⁶ cells/6 cm plate attach and multiply for 2 days but at about 50 h the cells enter a period of cell fusion to form a network of multinucleate fibres and cross-striations become obvious.

6.7. *Mouse embryo cultures*

– Kill pregnant (at least half-term) mice by breaking the neck.
– Immerse the mice in methylated spirits and place under a UV light for 5 min to sterilise them exteriorly.
– Aseptically, using forceps and scissors, cut through the skin under the forelegs and strip it off down to the rear legs.
– Using fresh forceps and scissors, open the abdomen and remove the uterus into a large Petri dish.
– Rinse each uterus with 25 ml of PBS-A containing 1% penicillin/ streptomycin.
– Remove the embryos with fresh forceps and scissors into warm PBS-A. Agitate and discard the PBS-A. Wash until clear. Check after each wash (using brain heart and Saboraud medium) that the embryos are contamination-free.
– Remove the embryos into sterile universals (about 6 embryos per bottle).

– Mince the embryos with scissors and wash with PBS-A until clear. Check for contamination as before after each wash.
– Add warm 0.25% trypsin in PBS-A (about 5 ml per embryo) to the mince and transfer to a 500 ml dimpled flask. Add a sterile teflon or silicone rubber covered magnetic stirrer.
– Stir *gently* at 37°C for 15 min.
– Remove from the magnetic stirrer and allow the tissue to settle. Remove trypsinised cell suspension, i.e. non-settled material, to a 250 ml centrifuge bottle containing 20 ml calf serum.
– Add fresh trypsin to the settled material. Repeat the trypsinisation once or twice, checking each suspension for contamination.
– Centrifuge the cell suspensions at 150 g for 5 min and discard the supernatant fluids.
– Resuspend the cells in a small volume of warm PBS and then fill the bottles with PBS.
– Allow large particles to settle to the bottom of the bottle. Pipette off any froth and floating particles.
– Filter the cell suspension through a sterile gauze filter to remove any clumps.
– Centrifuge at 150 g for 5 min.
– Seed Roux bottles with 40–50 × 10⁶ cells in 50 ml ETC (Appendix 6).
– Seed 80 oz. winchesters with 2 × 10⁸ cells in 200 ml ETC (Appendix 6).
– Gas with 5% CO_2 (see § 3.1.2) and incubate at 37°C for 4–5 days. Then check for sterility.
– Subculture by splitting 1 to 2 or 1 to 3.

6.8.   *Chick embryo cells*

– After a 10-day incubation period remove the eggs from the incubator and candle to determine viable embryos. Only well-developed active embryos with good blood supply are used.
– Leave under UV light for 10 min with the air sac end uppermost. Swab the entire egg with 70% alcohol.

– Crack the shell over the air sac and remove pieces of shell with sterile forceps.

– Cut away the shell membrane and chorioallantoic membrane using sterile forceps.

– With a pair of curved forceps, remove the embryo by the neck into a 90 mm Petri dish.

– Remove head (internal organs, e.g. liver may also be removed and processed separately; see below) and transfer the carcass to a conical flask (100 ml) of warm PBS-A (Appendix 1).

– Use fresh forceps and scissors for each egg or carefully sterilise the instruments by flaming.

– Swirl the embryos in PBS-A to wash off adherent blood, and discard PBS-A.

– Mince the embryos as finely as possible by chopping with scissors either in the conical flask or in a universal container (mince may be transferred in PBS-A if necessary).

– Add 10 ml 0.25% trypsin in PBS-A per embryo; add a sterile teflon or silicone rubber covered magnetic stirrer to the flask and stir at low speed (1 turn/s) for 10 min.

– Allow large clumps to settle and remove the turbid supernatant into a centrifuge bottle containing 20 ml calf serum. Add a fresh volume of trypsin to the tissue.

– Repeat the latter two steps once or twice.

– If only a few embryos are being trypsinised then the procedure may be carried out by gently shaking the mince and trypsin in a stoppered bottle in a water bath at 37°C for 0.5–1 h.

– Centrifuge the cells at 100 g for 5 min and discard the supernatant.

– Resuspend the cells in 20 ml warm growth medium (e.g. medium 199 containing tryptose phosphate (2%) and calf serum (2%). Allow the large clumps to settle and transfer cell suspension to a universal container.

– Dilute a sample from the cell suspension 1/10 in PBS-A containing 1% acetic acid and count.

  In a good preparation, 10^8 cells will be obtained from each embryo.

– Seed 8×10^6 cells into 90 mm Petri dishes.

Seed 20–30 × 10^6 cells into Roux bottles.
Only about half the seeded embryo cells attach and grow. If the cell concentration is lowered, the fraction which attaches decreases particularly at seedings of less than 1 × 10^6/90 mm dish (10^5 cells/ml).

- Incubate cultures at 37°C and change the medium on the following day.
- Subculture between the third and seventh day after explantation by dividing 1 to 4 ensuring not less than 10^5 cells/50 mm dish, and 5 × 10^5/90 mm dish.

Chick embryo cells are difficult to sustain for more than four or five subcultures without undergoing a marked reduction in growth rate.

6.9. Chick embryo liver cells

- Remove the embryo as described above.
- Using forceps and larger scissors remove liver aseptically and immerse in PBS-A in a weighed Petri dish. Repeat until approximately 0.8–1.0 g of tissue has been obtained.
- Cut into small pieces using fine scissors and forceps.
- Discard the medium and replace with 10 ml 0.25% trypsin in PBS-A at 37°C.
  Incubate for 15 min at 37°C in a humidified incubator (in an incubator containing a tray of water so as to maintain high humidity and thereby prevent evaporation of water from growth media, etc.).
- Discard the trypsin solution and wash the tissue twice with fresh PBS-A at room temperature in a bijou bottle.
- Immediately dissociate the tissue in a small volume of PBS-A by repeated aspiration. Do this by drawing the suspension into a Pasteur pipette or syringe and expelling it 10 times.
- Inject the concentrated cell suspension into 10 ml EC_{10} (Appendices 1 and 6) and repeat the aspiration a further 10 times.

- Allow undissociated tissue fragments to sediment and remove them.
- Count a sample of the cell suspension, and adjust the volume of the suspension (with EC_{10} – Appendices 6 and 1 – at 37°C) to give a cell density of around 3×10^6 cells/ml.
- Transfer 5 ml aliquots of the cell suspension to 5 cm Petri dishes and incubate at 37°C in a humidified CO_2 incubator.
- After 1 and 3 days remove 2.5 ml medium from each dish and replace with 2.5 ml fresh EC_{10} (Appendices 6 and 1).

During the first week of growth these primary liver cell monolayers show differentiative changes typical of chick liver at the time of hatching, e.g. chick liver UDP-glycuronyltransferase increases in specific activity up to 10-fold within a day of hatching and similar changes are reproduced in the monolayer (Skea and Nemeth, 1969).

6.10. Dipteran cell culture (Dolfini, 1971)

Horikawa and Fox (1964) isolated single cells from *Drosophila* embryos and were able to propagate these with the diploid chromosome number for years.

- Remove the chorion from blastodermal embryos by treatment with 3% NaClO for 2 min.
- Sterilise by treating for 15 min with 0.05% $HgCl_2$ in 70% ethanol.
- Dissociate the developing eggs by gently homogenising in a glass homogeniser and filter off the vitelline membranes using a 170–220 μm filter.
- Wash the cells in growth medium (pH 6.5) supplemented with 10% newborn calf serum.

Several cell lines have been isolated (Echalier and Ohanessian, 1969) and one of these has the following phase times (Dolfini et al., 1970):

G1   1.8 h
S   10.0 h

G2   7.2 h

T   18.8 h

Cell lines of *Aedes aegypti* (Grace, 1960) and *A. albopictus* (Singh, 1967) have been isolated as follows:

- Sterilise eggs as above and allow to hatch under aseptic conditions.
- Chop freshly hatched embryos and treat with 0.25% trypsin for 10 min at 37°C dispersing by pipetting.
- Wash the cells and incubate at 28°C in the medium of Mitsuhashi and Maramorosch (1964) supplemented with 10–20% foetal bovine serum at pH 7.0–7.4 to which non-essential amino acids and glutamine are added.

In this medium cells of *A. albopictus* will double every 10–11 h and have the following cell cycle phase times calculated using the graphical analysis method (Fig. 10.12).

G1 3.0 h

S   4.9 h

G2 1.5 h

M   1.1 h

These figures are quite similar to those found in vivo for brain cells of *A. aegypti* (Marchi and Rai, 1978).

CHAPTER 7

Cell culture media

7.1. Introduction

It is no longer common for laboratories to make up media from their individual constituents. Even when a special medium devoid of one particular amino acid is required it can usually be found in a commercial suppliers catalogue. The advantage of using commercially available media far outweigh their increased cost to all except those involved in studying the media themselves. For this reason detailed methods of media preparation from individual components are not given here. If required these may be found in Morton (1970) or in the original references. It cannot be too strongly stressed, however, that when preparing media water must first be deionised (always check the cartridge – see § 4.1.1) and then glass distilled at least once and stored in a glass or plastic container.

In the early work, embryo and tissue extracts were commonly used to encourage growth of cells but now these are only occasionally added to medium and then only for cultivation of certain primary cells (e.g. see § 2.1 and § 2.2). Today's media, however, as well as containing amino acids, vitamins, salts and glucose normally include serum at about 10% and sometimes also other additives such as Bactopeptone or tryptose phosphate (Appendix 1).

Starting with medium 199 (Morgan et al., 1950) which contains over 60 synthetic ingredients (Appendix 1: Table 18), many media formulations are available commercially. Unfortunately, the details of the methods of preparation and the absolute amounts of the various ingredients tend to vary from supplier to supplier. Morton

(1970) has surveyed a number of commercially available media listing the initial formulation and preparation method.

In 1955 Eagle identified the growth requirements for several mammalian cell lines in both qualitative and quantitative terms. His basal medium (BME – Appendix 1: Table 12) has 28 ingredients and is supplemented with 5% bovine serum. He later modified this medium to enable cells to grow without daily treatment and his minimum essential medium (MEM) is widely used today (Eagle, 1959), although further modifications have been suggested by other workers, e.g. Glasgow MEM, which is essentially twice the basal medium (BME) concentration of amino acids and vitamins with extra glucose and bicarbonate, has been recommended for the growth of BHK21/C13 hamster cells (McPherson and Stoker, 1962), and Dulbecco's modification, which contains four times the BME level of amino acids and vitamins together with some non-essential amino acids and added ferric nitrate, is recommended for propagating polyoma virus in hamster cells (Dulbecco and Freeman, 1969) (see Appendix 1: Table 12).

7.2. Balanced salt solutions

Media are based on a balanced salt solution (BSS) which, as well as supplying essential ions, is also important to maintain the osmotic balance. In addition, glucose is sometimes included in balanced salt solutions. However, it is better to add a sterile glucose solution along with filter sterilised sodium bicarbonate after the other ingredients have been sterilised by autoclaving. Salt solutions have developed over the years, but Earle's BSS and Hanks' BSS together with one devised by Eagle for suspension cultures are those in common use today. These are usually made up as tenfold (10×) concentrates without bicarbonate in glass distilled water and they can be stored in the cold room with a drop of chloroform in the bottom of the bottle to prevent bacterial growth (Appendix 1: Table 1). The stock solution is diluted with distilled water and autoclaved, after which bicarbonate and glucose may be added. The details of how to make up Earle's and Hanks' BSS are in Appendix 1 (Table 1).

The glucose and bicarbonate solutions are sterilised separately by filtration, and details of their preparation are given in Appendix 1 (Tables 3 and 4).

It is very important to maintain the correct pH of growth media between 7.2 and 7.5 and this is generally achieved with a bicarbonate/ CO_2 system. In the presence of 5% CO_2 in air this produces a pH of 7.4

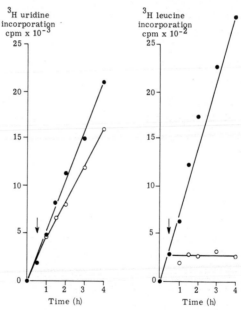

Fig. 7.1. Effect of increased NaCl concentration of the incorporation of [³H]uridine and [³H]leucine into CHO cells. Hamster CHO cells were set up as coverslip cultures in a multiwell dish in Eagle's medium (Glasgow modification) containing 10% calf serum (10^5 cells/0.5 ml). After overnight growth, 10 µl of [³H]uridine (0.5 µCi) or [³H]leucine (2 µCi) was added per well, and 30 min later 10 µl 5 M NaCl was added to half the wells. Coverslips were harvested at the indicated times and washed twice in ice-cold BSS, 4 times in cold 5% TCA and twice in absolute ethanol. The cells were dissolved in 0.5 ml hyamine hydroxide and the radioactivity counted using a scintillator of 0.5% diphenyloxazole in toluene. ●, Control, i.e. 120 mM NaCl; ○, 220 mM NaCl.

(the phenol red imparts a tomato colour to the medium at this pH; if the medium becomes too alkaline it turns red and then puce while when it is too acid it becomes yellow). As growing cells produce acid it is sometimes necessary to increase the level of bicarbonate in the medium in an attempt to maintain pH. The phosphate in BSS also helps to maintain the pH as well as supply essential phosphate ions.

As well as being essential for certain reactions Na^+ and, to a lesser extent, K^+ are important in maintaining the osmotic balance in the cell. The sodium ion concentration is maintained about isotonic largely by the NaCl in the BSS but in Earle's BSS this is supplemented with $NaHCO_3$ and to a lesser extent with other salts (see Appendix 1). If the Na^+ concentration is increased from the normal 120 mM up to 220 mM it leads to a dissociation of polysomes with concomitant inhibition of protein synthesis and cell growth (Fig. 7.1) (Saborio et al., 1974).

Calcium ions are essential for the attachment of cells to a glass or plastic surface and hence are omitted from BSS for use in suspension cultures and usually the phosphate concentration is raised tenfold (Eagle, 1959).

Dulbecco's phosphate buffered saline (PBS) is similar to Hanks' BSS but bicarbonate is omitted and the levels of Na_2HPO_4 and KH_2PO_4 are raised to provide increased buffering capacity. It is not used as a basis for growth medium but is often used for washing cell monolayers. PBS is made up in three parts (Appendix 1: Table 2). PBS solution A (PBS-A) lacks the Ca^{2+} and Mg^{2+} ions which are added separately as PBS solution B and PBS solution C, respectively, to constitute PBS proper.

7.2.1. Zwitterionic buffers

Hepes has been shown to be non-toxic to cells and can be used instead of bicarbonate in which case cells need not be maintained in an atmosphere of 5% CO_2 in air. However, as bicarbonate is essential for cloning cells (see § 8.1.1) mixtures of Hepes and bicarbonate are used and the cells grown in an atmosphere of 2% CO_2 in air.

Hepes is 4-(2-hydroxyethyl)-1-piperazineethane sulphonic acid.

$$\text{HO·CH}_2\text{·CH}_2-^+\text{NH} \bigcirc \text{N-CH}_2\text{-CH}_2\text{SO}_3' \qquad \text{MW} = 238.3$$

At 37 °C is has $pK_{a_2} = 3$ and $pK_{a_2} = 7.31$. It is thus a much more suitable buffer than bicarbonate which has a $pK = 6.10$.

Hepes is used at 10–25 mM and is added to medium from a stock solution (1 M) whose preparation is described in Appendix 1, Table A1.5. As well as Hepes, other zwitterionic buffers have been used in cell culture medium. TRICINE (N-[(Tris-hydroxymethyl)-methyl] glycine, $pK_a = 7.79$ at 37 °C) has been used in Eagle's MEM (Spendlove et al., 1971) and in Swim's 577 (Gardner, 1969) and TES, (N-[(Tris-hydroxymethyl)methyl]-2-aminoethanesulphonic acid, $pK_a = 7.16$ at 37 °C) has been used in Eagle's MEM, BME and in medium 199 (Williamson and Cox, 1968; Massie et al., 1972). These buffers are used in varying concentrations (10–50 mM), either alone or more often in combination with bicarbonate at 5–15 mM. Eagle (1971) suggests combinations of buffers which can be used to buffer the medium over the range 6.4–8.35.

7.3. Eagle's medium

This, in its various forms (e.g. BME, MEM; see Appendix 1; Table 12), is generally suitable for the cultivation of most cell lines for which it is supplemented with 10% calf or foetal calf serum and occasionally with tryptose phosphate.

Basal medium (BME) needs to be changed at least every other day to support continued cell growth while Minimum Essential Medium (MEM) supports growth for several days.

Eagle's media contain, in addition to the salts and glucose of BSS, 12 essential amino acids and 9 vitamins. Glutamine and, if required, antibiotics along with serum etc., have to be added also before use.

It is common practice when culturing certain cells, e.g. BSC1 and CV1 monkey cells, to supplement Eagles MEM with certain non-essential amino acids (Appendix 1).

The different forms of Eagle's medium are available commercially. For example from Flow Laboratories or Gibco Biocult (Appendix 3). Complete 1× MEM medium only requires the addition of glutamine, antibiotics and serum, and this is very suitable for small scale work in laboratories which lack facilities for preparing media. It is, however, a very expensive way of buying media if more than a litre or so is required per month.

10× MEM medium simply requires dilution into sterile distilled water followed by addition of glutamine, serum, bicarbonate and antibiotics, if required.

50 and 100× concentrates of vitamins and amino acids may either be added to the appropriate volume of sterile BSS and glutamine and serum added, or they may be mixed to make a 10 × MEM stock (Appendix 1: Tables 13 and 14). This we have found a very suitable way of making medium as the glucose and glutamine may be incorporated into the 10× MEM stock. It is significantly cheaper than buying Eagle's 10× MEM, yet occupies much less cold room space than if 1× medium were reconstituted initially. The methods of preparation of 10× stocks of Eagle's MEM and Dulbecco's MEM are indicated in Appendix 1 (Tables 13 and 14). These methods involve filtration of the 10× MEM, but this can be avoided if the glucose and glutamine are dissolved separately and sterilised by filtration and the reconstitution is carried out aseptically using sterile distilled water.

Growth media, i.e. media suitable for the continued culture of established cell lines, are prepared by dilution of the 10× stock solutions. If the 10× stock is made with salts present (i.e. if it is obtained commercially or if it is made from powder – see below) then it must be diluted with water. If, however, it is made as described in Appendix 1, Table 13 or 14, it must be diluted with BSS as indicated in Table 15. The latter table gives methods of preparing various 'minimal essential media' supplemented with calf or foetal calf serum and in one case also with 'non-essential amino acids'. The various media are all based on Eagle's MEM and have been developed to improve the growth of certain cell lines or strains as indicated.

7.3.1. Powdered media

Recently, powdered media have become available commercially from for example Flow Laboratories or Gibco Biocult Laboratories (see Appendix 3). This is the cheapest way of buying media if large quantities are used; the powder occupies little space and may be stored in the cold room in sealed containers for 6–12 months.

Powdered medium should be reconstituted at room temperature as per manufacturers instructions, which leads to 1× medium. It is important not to leave traces of powder undissolved. Bicarbonate is added separately and the pH adjusted to about 7.1 before sterilisation by filtration. Before use, glutamine should be added, if not already present, together with serum (usually to 10%, as described in Appendix 1).

Each batch should be checked for bacterial contamination in a) Saboraud fluid medium at 31°C for 1 week (Appendix 4); and b) brain heart infusion broth at 37 °C for 1 week (Appendix 4).

Autoclavable powdered medium, where the phosphate buffer has been replaced with a succinate buffer (Yamane et al., 1968), should be dissolved in 95% of the final volume of distilled water and the pH adjusted to 4.1 before autoclaving for 15 min at 121 °C. After cooling to room temperature 3 ml sterile 7.5% $NaHCO_3$ is added per 95 ml medium, together with glutamine and serum etc., and the pH is adjusted to 7.2–7.4 with sterile 1 N NaOH, if necessary.

Although not generally recommended by the manufacturer, if care is taken to ensure that all the powder is dissolved we have found that Glasgow MEM powder can be reconstituted to give a 10 × stock which then only requires dilution with sterile distilled water and pH adjustment before use. In this case it is important not to add bicarbonate or adjust the pH until the medium has been diluted.

The advantages of reconstituting to 10× are: (1) there is a smaller volume to filter sterilise; (2) the 10× stock occupies less cold room space; (3) it is more economical to buy bottles which reconstitute to 5 l of 10 × stock (50 l of 1 × medium) than smaller amounts.

7.4. More complicated media

Different media have been developed for specific cell lines in order to obtain optimal growth or in an attempt to grow cells in defined media without the addition of serum.

Thus McCoy's medium 5A (McCoy et al., 1959) has been used as a standard medium for cloning cells. It is based on BME, amino acids and the vitamins from medium 199 (Appendix 1: Table 18). This was modified further to form RPMI-1629 (Appendix 1: Table 17) for long-term culture of leukaemic myoblasts (Armstrong, 1966).

Ham's F10 medium was very carefully formulated for cloning diploid hamster ovary cells (Ham, 1963) and was later modified – Ham's F12 (Appendix 1: Table 16) (Ham, 1965) – to support growth with addition of minimal amounts of serum.

The Connaught Medical Research Laboratories Media are also designed to support cell growth without the addition of serum. Thus CMRL 1066 (Appendix 1: Table 20) will support the growth of several cell lines as will the National Cancer Institute medium NCTC 135 (Appendix 1: Table 19) and CMRL 1415 comes nearest to supporting the growth of cell strains (Healy and Parker, 1966a,b; Evans et al., 1964).

Many of these more complex media are available from commercial suppliers, e.g. Flow Laboratories or Gibco Biocult (see Appendix 3) at 1 × formulation, and the constitutions of some of them are given in Appendix 1.

7.5. Simple media with unspecified additives

There are a number of media available which are not based on a detailed investigation of growth requirements, but rather include lactalbumin hydrolysate (Appendix 1: Table 9) or yeast extract (Appendix 4) to provide an inexpensive source of amino acids or vitamins. Thus Melnick's monkey kidney media A and B (Melnick, 1955) contain lactalbumin hydrolysate and calf serum in Hanks' and Earle's BSS, respectively. Mitsuhashi and Maramorosch mosquito cell medium contains lactalbumin hydrolysate, yeast extract and foetal calf serum in a specially developed saline (Mitsuhashi and Maramorosch, 1964; Singh, 1967).

7.6. Antibiotics

The method of preparation of an antibiotic stock solution for use in cell culture media is given in Appendix 1: Table 6. Penicillin at 0.2–1.5 U/ml inhibits the growth of most gram-positive bacteria. Streptomycin sulphate at 100 µg/ml effectively suppresses growth of gram-negative bacteria, but it has a half-life of only 4 days under physiological conditions. Dihydrostreptomycin sulphate is more stable but either may be used. Vanomycin at 100 µg/ml is active against gramnegative bacteria and also mycoplasma, and it can be used to rid cultures of mycoplasma (see Chapter 9).

The use of antibiotics in cell cultures should be restricted to short-term cultures, and routine subculturing of cell lines should be done in the absence of antibiotics so that selection for antibiotic-resistant strains of bacteria is discouraged. Further information on antibiotics may be found in Chapter 9.

7.7. Serum

Despite attempts to grow cells in fully synthetic media, the vast majority of cells are grown in media containing some natural additive, usually serum. Although a variety of sera are commercially available, most cells grow best in the presence of bovine serum. This is obtained either from the foetus or the newborn calf (foetal bovine or calf serum). Obtaining serum (especially foetal bovine serum), processing and sterilising it is a difficult procedure requiring special apparatus and is best left to the commercial companies, e.g. Flow Laboratories, Gibco Biocult, etc. (see Appendix 3).

It is important that haemolysis is kept to a minimum and only suitable batches of serum are processed. Processing involves passage through a series of filters down to 0.1 µm, after which the product undergoes a series of tests to detect possible bacterial, mycoplasmal or viral contamination. In addition, each batch of serum is tested to ensure that it will support the growth of primary, diploid and established cell lines, using both serial passage and plating efficiency tests.

Before buying serum (which is an expensive item) the customer is recommended to obtain samples of various batches and to repeat a plating efficiency test (§ 8.1.2) or to measure the growth rates and yields of at least two cell types. It is particularly important that the batch of serum supports the growth of the cells in the customer's laboratory. Most commercial companies are in favour of this trial and moreover will retain supplies of a particular batch of serum in their own – 20 °C store, delivering a fraction of the order each month. Once prepared, serum should be stored at – 20 °C at which temperature it remains stable for 6 months or more. It should be thawed slowly (by standing at 4 °C) and not be subjected to repeated cycles of freezing and thawing.

7.7.1. Removal of small molecules from serum

This is essential for nutritional studies and is often desirable when labelling cultures with certain radioisotopes, so that (a) the specific activity of the precursor may be accurately known, and (b) the specific activity may be the maximum possible. Thus, it is common to label cells with a tritiated amino acid to follow the rate of protein synthesis, and so as to maximise incorporation, special medium is made lacking that amino acid. Serum may contain, however, a significant amount of that amino acid and the best results will only be obtained if this is first removed. Two methods are available.

1. *Sephadex treatment.* A sterile column (2.5 × 100 cm) containing about 50 g Sephadex G-50 (coarse) is well washed with 5 l sterile 0.85% saline. The Sephadex slurry may be sterilised by autoclaving for 15 min at 121 °C before packing. 200 ml of whole serum may be applied to such a column and eluted with saline at 4–5 ml/min. The excluded fraction elutes between 200 and 400 ml.

2. *Dialysis.* Dialysis tubing, sterilised by boiling in distilled water, may be filled with serum, using aseptic conditions, and dialysed against 2 changes of 100 volumes of sterile saline or BSS at 4 °C over 48 h. At the end of this time the serum may be removed from the sac again using aseptic conditions.

Table 7.1 is taken from Patterson and Maxwell (1973) and gives the composition of serum before and after treatment.

Only limited proteolysis occurs over a 3-week period even at 37 °C,

TABLE 7.1

Material removed from serum by gel filtration or dialysis

Constituent	Whole serum	Sephadexed serum	Dialysed serum
Alanine (mM)	0.672	0	< 0.001
Arginine (mM)	0.246	< 0.001	0
Aspartic acid (mM)	0.146	< 0.001	< 0.001
Glycine (mM)	0.408	< 0.001	< 0.001
Glutamic acid (mM)	0.309	0	< 0.001
Histidine (mM)	0.216	< 0.001	< 0.001
Isoleucine (mM)	0.121	< 0.001	0
Leucine (mM)	0.347	0	0
Lysine (mM)	0.129	0	< 0.001
Methionine (mM)	0.015	0	0.005
Phenylalanine (mM)	0.126	0	0.001
Proline (mM)	0.387	< 0.001	0
Serine (mM)	0.295	< 0.001	< 0.001
Threonine (mM)	0.261	< 0.001	< 0.001
Tyrosine (mM)	0.498	0	0
Valine (mM)	0.336	0	0
Total protein (g%)	7.3	6.2	6.5
Cholesterol (mg%)	135	113	122
Uric acid (mg%)	1.4	0.8	0.7
Glucose (mg%)	132	6	7
Potassium (mEq/l)	5.8	0	0
Calcium (mg%)	7.2	2.8	0.2
Alkaline phosphatase (units)	15.9	13.2	14.0

but it is not recommended that such serum be stored for long times, otherwise small molecules may again be released.

7.7.2. Role of serum

Ham and McKeehan (1978) have recently reported progress in their attempts to grow diploid cells in serum-free medium. They point out that precise quantitative adjustment of nutrient concentrations is important and that the nature of the culture surface may have a pronounced effect on clonal growth of cells. They were able to reduce the concentration of dialysed serum to 500 µg protein/ml.

Although certain factors have been isolated from serum, which have been shown to be essential for cell growth, there is still controversy about how many and what sort of factors are present. Thus it may be that hormones bound to the serum proteins are essential to stimulate certain cells to grow (Leffert et al., 1977; Hayashi et al., 1978). Hayashi and Sato (1976) have replaced serum by five hormones (triiodothyronine, thyrotropin releasing hormone, transferrin, somatomedin and a biologically active peptide of parathyroid hormone) and obtain equivalent growth rates for GH_3 cells (a rat pituitary cell line which produces growth hormone and prolactin) in Ham's F12 medium. A mixture of 25 hormones can replace serum for growth of BHK and HeLa cells, but not all these have been shown to be essential.

Fibroblast growth factors and platelet derived factors are polypeptides which have been purified from serum and spent medium (Burk, 1976; Antionades and Scher, 1977; Ross and Vogel, 1978). They may be of upmost importance for growth of fibroblasts and glial cells (Westermark and Wasteson, 1976; Rutherford and Ross, 1976; Vogel et al., 1978). Westall et al. (1978) have shown that a basic protein released from myelin breaks down to give two smaller fibroblast growth factors, and Calissano et al. (1978) have shown that nerve growth factors can organise actin and activate ATPases perhaps acting by a stimulation of cyclic AMP production (Raff et al., 1978; Green, 1978). On the other hand, Rudland et al. (1974) claim that fibroblast growth factor activates guanyl cyclase, and Moenz et al. (1975) have shown that SV40 transformed 3T3 cells have higher cGMP levels than untransformed 3T3s. Miller et al. (1975) have reviewed the arguments for and against the role of cyclic nucleotides in the control of cell growth.

Epidermal growth factors (Richman et al., 1976) and multiplication stimulating activities (Bolen and Smith, 1977) have also been purified. Epidermal growth factor (EGF) and insulin bind to cell membranes forming receptor complexes which are then taken into the cell and become associated with the perinuclear region (Das and Fox 1978; Haigler et al., 1978; Schlessinger et al., 1978). The membrane proteins of the complex include glycoproteins (Maturo and Hollenberg, 1978; Sahyoun et al., 1978) and the action of EGF may be to promote their phosphorylation (Carpenter et al., 1978). Although internalisation of fluorescent labelled EGF occurs (Haigler et al., 1978), this may not be relevant as Schechter et al. (1978) have shown that some EGF remains tightly bound to a few cell surface receptors up to 8 h after a 30 min exposure of cells to EGF. Even at this time treatment of cells with an antibody specific for EGF still abolishes its biological action of stimulating DNA synthesis and cell division.

Baker et al. (1979) and Linsley et al. (1979) have identified a 185,000 molecular weight receptor protein to which EGF binds. Glenn and Cunningham (1979) have suggested that the mitogenic action of thrombin on non-dividing chick embryo, mouse embryo and human foreskin cells maintained in serum-free medium is a result of binding to and cleavage of a 43,000 molecular weight cell surface receptor protein. It must be emphasised, however, that there is a reciprocal relation between proliferation and differentiation, and in cultured adrenocortical cells ACTH, which stimulates steroidogenesis, inhibits cell proliferation while fibroblast growth factor has the reverse effect (Gill et al., 1978).

7.8. Other natural additives

To make up for undefined limitations in growth media crude mixtures of nutrients are sometimes added to promote cell growth. The methods for preparation of lactalbumin hydrolysate, chick embryo extract and tryptose phosphate broth are given in Appendix 1: Tables 9, 10 and 11, and their use is referred to where appropriate throughout the book.

7.9. Media for culture of insect cells

The aim in designing a culture medium is to mimic to some degree the composition of the fluid surrounding the cells in vivo. This proved difficult for insect cells in the 1930s as the composition of haemolymph was unknown and has been shown more recently to be less important than originally considered (Yunker et al., 1967; Vaughan, 1971). Thus, although the inorganic salt concentration of insect culture medium may reflect that of haemolymph the amino acid and vitamin component may be supplied by lactalbumin hydrolysate, whole egg ultrafiltrate and bovine serum when haemolymph is not available. Foetal bovine serum proved to be the most effective of several sera tested (Mitsuhashi and Maramorosch, 1964; Singh, 1967). The high levels of amino acids found in haemolymph may be supplied from a 10% lactalbumin hydrolysate supplemented with non-essential amino acids. Few studies have been done on the amino acid requirements of cultured insect cells and, although some media (Grace, 1962) do specify an amino acid mixture, the presence of additives, such as whole egg ultrafiltrate, makes the significance of specific amino acid concentrations doubtful.

Glucose is the usual carbohydrate for energy source in culture even though in vivo a variety of carbohydrates are found in haemolymph (Wyatt and Kalf, 1957). The B vitamins and ascorbic acid are considered essential (Samborn and Haskell, 1961).

The inorganic constituents vary widely between one insect culture medium and another. The pH is generally just below 7, but even this is not always optimal as *Aedes albopictus* cells are cultured at pH 7.2.

CHAPTER 8

Techniques

8.1. Cell cloning and plating efficiency

Due to inadequacies in the early growth media the first successful cloning of a somatic cell was not achieved until 1948 (Sandford et al., 1948).

The problems that arise with cloning are a result of the attempts to grow cells at very high dilutions. In this case the medium must provide nutrients which may be the products of cell metabolism, but which escape from the cells before they can be utilised in subsequent biosynthetic reactions. Cells growing in culture constantly lose nutrients to the medium even while they remove nutrients from the medium for growth. For cells growing at concentrations of 10^5–10^6/ml the medium quickly becomes *conditioned*. This means that the medium now contains a number of components released into it from the growing cells. These components may be simple amino acids, e.g. glycine, not normally present in simple media (§ 2.2) or more complex growth factors (§ 7.7.2). However, when the cell concentration is reduced to 1 or 2 cells/ml the outflow of, for instance, glycine cannot be balanced by an equivalent uptake and cells die of starvation. The requirements of cells growing at low density include pyruvate and non-essential amino acids and also carbon dioxide (Eagle, 1971). It is therefore important either to clone cells using a bicarbonate/CO_2 buffer or, if organic buffers such as Hepes are used, they must be supplemented with bicarbonate and CO_2 (e.g. 20–25 mM Hepes plus 8 mM bicarbonate and 2% CO_2 in air gas phase). Two approaches have been used to circumvent this problem. Cells are either cloned in a restricted microenvironment, e.g. a capillary tube in which they can rapidly condition the medium, or they are cloned in the presence of a large number of killed cells, i.e. a feeder layer irradiated with X-rays (Puck et al., 1956; Sandford et al., 1961). Ham's medium F12 has been designed specifically for cloning cells (Ham, 1965) and should be supplemented with 10–30% foetal bovine serum. Great care must be taken to prevent the medium becoming too alkaline by loss of CO_2. These methods are, however, largely unnecessary when it is desired to clone continuous cell lines. Thus a dilute suspension of HeLa cells (containing about 100 cells in 5 ml) will grow up to form colonies, each derived from a single cell. It is important to prevent the movement of these cells (and hence mixing of clones) and sometimes they are overlayered with soft agar.

The clones may be isolated either by surrounding each colony with a stainless steel cylinder and removing the cells by trypsin action or by initially seeding the cells into a dish containing numerous coverslip fragments and selecting those fragments on which single colonies become established.

For measuring plating efficiency a similar procedure is adopted and the colonies are fixed and stained (with Giemsa) and their number compared with the initial input. The plating efficiency of primary cells is low, but plating efficiencies of up to 100% can be achieved with cell lines.

8.1.1. Simple methods for cloning or measuring plating efficiency

– Wash the cell monolayer in BSS to remove serum.
– Release cells with 0.25% trypsin in PBS or BSS (the divalent ions improve plating efficiency) and disperse by gently pipetting up and down the minimum number of times. Trypsinisation at 4°C for 2–10 min has been recommended especially for cloning cells in low serum.
– Add an equal volume of complete medium. Medium F-12 supplemented by up to 30% foetal calf serum has been recommended (Ham, 1965). The medium should be preincubated with 5% CO_2 to prevent undue alkalinity and should not be above 30°C, otherwise

the cells will attach to the walls of the vessel (Ham and Puck, 1967).

- Dilute so that 5 ml contains between 50 and 250 cells and quickly inoculate into a 5 or 6 cm dish (see cloning, method 1 below).
- Incubate for 8–15 days at 37°C in a humidified CO_2 incubator (§ 6.9).

If the cells migrate quickly intermixing of colonies may occur. To prevent this, medium containing 0.17% agar may be used (§ 8.1.4).

For plating efficiency
- Fix the resulting clones in 25% formalin and stain with Giemsa (Appendix 2) to estimate plating efficiency.

For cloning

Method 1
- If clones are to be isolated the culture vessel should contain many small fragments (0.05 cm² of coverslips to which the cells will attach and grow. Using sterile forceps remove coverslip fragments on which single cells have attached (check this microscopically) and transfer them to separate wells in a tissue culture tray (§ 3.2.1).
- Add 0.3 ml complete medium to each well and continue the incubation until the clone covers the surface of the well. The cloned cells may now be harvested by trypsinisation (§ 5.2) and transferred to appropriate vessels.

Method 2
- Clones may be isolated from 5 cm dishes using stainless steel cloning cylinders (about 6 mm in diameter and 12 mm high (Dow Corning; Appendix 3). These, and some silicone grease should be sterilised by autoclaving.
- Mark the position of a well isolated colony of cells using a grease pencil or magic marker on the undersurface of the dish.
- Remove the medium and rinse with BSS (Appendix 1).
- Using sterile forceps touch one end of a cloning cylinder into the silicone grease and place it over the colony to be isolated.
- Using a syringe and a 20 gauge needle add a drop of 0.05%

trypsin in BSS to the cylinder to loosen the cells which should then be withdrawn back into the syringe and transferred to a fresh Petri dish.

Method 3
– For cloning, the dilute cell suspension may be plated out on special Petri dishes whose growth surface is made of a membrane (Petriperm: Appendix 3).

This membrane allows ready diffusion of gases in and out of the dish from below and is also readily cut into pieces. In this way pieces containing single colonies may be cut out and transferred to the wells in a tissue culture tray (cf. broken coverslips).

8.1.2. Capillary method

The capillary method is rather difficult but is described in detail by Sanford (1973). Rather than growing the cells in the microenvironment of the capillary tube, Macpherson (1973) recommends their transfer in a microdrop of medium to the wells of multicup polystyrene trays, e.g. the 96 well trays available from Linbro Chemical Co. Inc., or Titertek Inc. (Appendix 3).

Cells are first released into suspension by very gentle trypsinisation (see above) and individual cells selected by sucking them up into micropipettes drawn from sterile soft glass tubing plugged at both ends. The drawing of the glass tubes should be in two stages. The first stage should yield two Pasteur pipettes of external tip diameter about 1 mm. In the second stage the narrow part of the pipette should be softened over the pilot flame of a bunsen burner and then removed and immediately drawn out to reduce the tip diameter to 0.1 mm or less.

The single cells are transferred to the multicup tray and each cup is examined microscopically to ensure it contains only one cell. About 0.1 ml medium is then added to each cup and the whole tray incubated in a humidified CO_2 incubator for 7–10 days until large colonies have grown. These may then be transferred to 5 cm dishes or small bottles.

An alternative method recommended by Cooper (1973) involves diluting a cell suspension to a concentration of 10 cells/ml. It is important that the cells are not stuck together in pairs or groups, i.e. it must be a single cell suspension. 0.1 aliquot is then dispensed into the wells of a microtitration plate. The wells are then immediately inspected with an inverted microscope and those containing only one cell are noted.

8.1.3. Cloning under agar

Under certain conditions cells will form colonies when not attached to a glass or plastic substratum but when suspended on soft agar. Macpherson (1973b) points out factors which promote the growth of cells in soft agar. These involve transformation with oncogenic viruses, infection with mycoplasma, increased levels of serum and addition of feeder cells or DEAE dextran to neutralise the acidic groups present in agar.

- A 1.25% solution of Difco Bacto-agar is made up in hot distilled water and sterilised by autoclaving. The agar sets on cooling and must be melted in a boiling water bath and cooled to 44°C before use.
- Add 20 ml calf or foetal calf serum and 20 ml tryptose phosphate broth (Appendix 1) to 80 ml double strength medium (e.g. Glasgow MEM; Appendix 1) and warm the mixture to 44°C.
- Add some of the melted agar to the above supplemented medium and mix gently to yield 0.5% agar medium.
- Pipette 7 ml 0.5% agar medium into 5 cm Petri dishes and allow to set at room temperature. Use within 1 h.
- Mix 1 volume of cell suspension with 2 volumes of 0.5% agar medium at 44°C and add 1.5 ml as an upper layer onto the agar base layer prepared above. Up to 10^3 colony forming cells should be present per dish.
- Incubate in a humidified CO_2 incubator for 7–10 days during which time colonies of 0.1–0.2 mm diameter form.
- Colonies may be removed from the agar into a tube containing 1 ml medium using a finely drawn pipette.

– Pipette the suspension up and down to break up the agar and release the cells into suspension.
– Transfer the cells to 5 cm dishes and add 5 ml agar-free medium. Incubate in a humidified CO_2 incubator (§ 6.9).

8.1.4. Cloning with a feeder layer

Confluent cultures containing 2×10^6 cells/90 mm dish are irradiated with 4000–6000 rads of γ-radiation from a cobalt source. Irradiation should last for less than 1 min and a variety of cell lines (e.g. HeLa, BHK21/C13 or 3T3) are suitable (Puck et al., 1956). These cells may be used directly or may be trypsinised into smaller vessels. They may appear healthy for up to 4 weeks but do not divide.

An alternative method of producing a feeder layer is to incubate the monolayer with mitomycin C (10^{-6} M) for 16 h which causes crosslinking of the DNA (Iyer and Szybalski, 1964). The monolayer is then washed three times with BSS to remove the mitomycin C and may be used directly or after subculture into smaller vessels. For cloning 6 cm dishes containing about 2×10^5 killed cells (i.e. irradiated or treated with mitomycin C) are used. These form the layer of feeder cells. Replace the medium in the dish of feeder cells with a suspension of the cells to be cloned. As few as 10 cells may be present but about 1000 is better. These cells settle into open spaces between the feeder cells and begin to grow and form colonies which may be isolated using cloning cylinders (§ 8.1.2).

8.2. Cell counting procedures

There are two methods of estimating the number of cells in a suspension. Using a haemocytometer the number of cells in a given volume is counted by direct microscopic examination. Using electronic cell counters, e.g. the Coulter counter, the cells in a given volume of suspension are drawn through an orifice and registered electronically.

8.2.1. Haemocytometer

Place the precision ground coverslip on the haemocytometer slide (Fig. 8.1) so as to cover the two ruled areas and press down gently until Newton's rings are visible. This leads to formation of a chamber of precise dimensions as the edges of the slide are raised exactly 0.1 mm above the ruled area. Each ruled area consists of 9 large squares, each of 1×1 mm, i.e. the volume above each large square is $1 \times 1 \times 0.1 = 0.1$ mm^3. (The 4 corner squares are subdivided into 16 smaller squares; the centre square into 25 smaller squares and the remaining squares into 20 smaller squares, but this is unimportant when counting cells.)

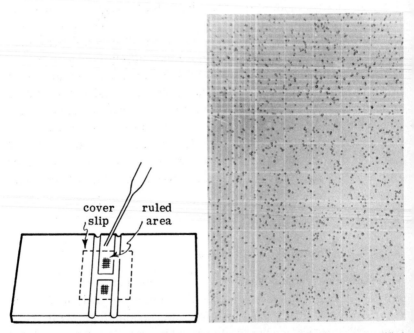

cover slip

ruled area

Fig. 8.1. Use of the haemocytometer. The diagram on the left shows a modified Neubauer haemocytometer and on the right is a photomicrograph of BHK21C13 cells ready for counting.

A drop of cell suspension (containing between 10^5 and 10^6 cells/ml) is placed at the two edges of the coverslip so that the suspension flows into the chambers by capillary action (Fig. 8.1). Do not flood the chamber. The slide is viewed under a low power objective of a microscope.

The cells in four large squares in the corners of each of the two ruled areas are counted. Count cells touching the right and top lines but not those touching the left and bottom lines. The volume of a large square is 0.1 mm^3 and the average cell count should therefore be multiplied by 10^4 to give the number of cells per ml.

8.2.2. Electronic cell counter

The principle of this method is that as a cell passes through an orifice it interrupts an electric current and this interruption is picked up as an impulse which is translated into a visual signal on an oscilloscope and recorded on a counter (Fig. 8.2). The size of the visual impulse is proportional to the volume of the cell and hence electronic 'gates' can be set to count cells of given sizes. The simpler machines (e.g. Coulter counter model D industrial) have only a lower threshold to the gate which ensures that particles of dust are not counted. The more complex machines (e.g. Coulter counter model ZB1) can be fitted with 'cell plotters' which automatically count cells in a series of gates and plot a histogram of cell number against cell size (Fig. 8.3).

The orifice through which the cells pass can vary in size but for animal cells it is 100 μm in diameter and 75 μm long and is manufactured from a ruby. The orifice tube is therefore fragile and expensive.

Prepare a counting saline solution (NaCl, 0.7%; citric acid, 1.05%; mercuric chloride, 0.1%) and to 24.5 ml in a small beaker add 0.5 ml of cell suspension (approx. 10^6 cells/ml); alternately 0.7 ml suspension may be added to 16.8 ml counting saline in a universal bottle). Mix the saline suspension well and set it on the counting table with the orifice tube and outer electrode immersed. Open the upper stopcock which connects the mercury manometer to a vacuum produced in a bottle by a small pump. This draws the mercury down the manometer.

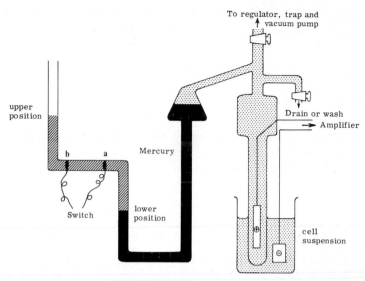

Fig. 8.2. Diagrammatic representation of an electronic cell counter. When the upper tap is open the vacuum pump draws the mercury into the position shown in black (lower position). On closing the tap the mercury slowly returns to its equilibrium (upper) position and while so doing it draws cell suspension through the small orifice. As a cell passes through the orifice it interrupts the current flowing between the two electrodes and each cell is registered on an oscilloscope and is counted. The count is recorded as the mercury sweeps out the volume between the switches (usually 0.5 ml).

After a few seconds the stopcock is closed and the mercury in the manometer returns to its original position. As it does so it draws cell suspension through the orifice. At the same time the mercury switches on the cell counter as it passes a wire electrode (a on Fig. 8.2) and switches it off again as it passes a second wire electrode (b on Fig. 8.2). Between the two electrodes the mercury sweeps out exactly 0.5 ml and hence draws this volume of saline cell suspension through the orifice.

In the standard procedure 0.5 ml of a 1/50 dilution (i.e. 0.01 ml) of the original suspension is counted but this may be varied depending on the concentration of the original suspension.

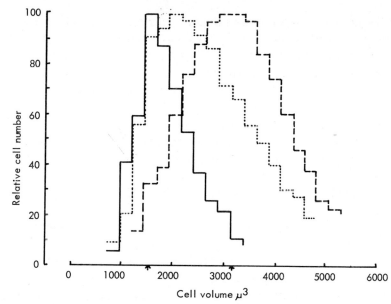

Fig. 8.3. Electronic cell volume plotter. Mouse L929 cells were harvested from stationary phase cultures at zero time and subcultured into medium to which thymidine (5 mM) was added at 8 h. At 24 h the thymidine was removed. At various times the cells were harvested by trypsinisation and their volumes measured using a Coulter counter model B fitted with a model J plotter. ———, 0 h; ————, 28 h; ...., 32 h. (Reproduced from Adams, 1969b, with the publisher's permission.) These models have now been replaced with the Coulter counter method ZB1 with the C1000 Channelyzer.

8.2.3 Comparison of the methods

The disadvantage of using the haemocytometer is that it becomes tedious for large numbers of samples. Moreover, there is an error of about 10% in estimations (Sandford et al., 1951). Some of the errors arise during dilution of the cell suspension which should be done with care. An advantage is that the apparatus required is quite cheap (Neubauer counting chambers can be obtained from Gelman Hawksley) and very small numbers of cells are required. It is thus easy to estimate the number of cells in a small dish by trypsinising them into a volume of 1 ml.

Because the cells are viewed directly, this method enables a judgement to be made about the quality of the cell suspension, i.e. clumped or broken cells present. By diluting the cell suspension with a solution of a vital dye, e.g. trypan blue at a final concentration of 0.2 g/l PBS-A (Appendix 1), and counting only the unstained cells a measure of the viable cell count is obtained. (N.B. Trypan blue is toxic and should not be allowed to come in contact with the skin.)

The electronic cell counter is, however, more reproducible. As many thousand cells are counted, machine sampling errors are very small and errors of less than 5% are easily achieved. The main source of error is in the initial sampling of the cell suspension which must be done with care. The number of cells in a small dish may also be estimated using an electronic counter but in this case almost all of the cell suspension is used in the estimation. If two or more cells pass through the orifice simultaneously they will be counted as one and corrections must be applied with cell concentration above $10^4/0.5$ ml. The coincidence error at this concentration is 3% but it rises to 22% at $10^5/0.5$ ml. One big advantage is the visual display on the oscilloscope of the range of cell sizes present in the suspension. On the one hand, this indicates if clumps of cells are present, and on the other it gives a measure of the distribution of cells around the cell cycle (see Chapter 10).

In addition, the use of gates or windows in the Coulter counter model ZB enables cell sizes to be accurately determined after the machine has been calibrated with pollen grains of known size. Thus ragweed pollen has a mean volume of 3800 μm^3. When coupled to a 'Channelyzer' a histogram of the cell size distribution is automatically recorded (Fig. 8.3). Alternatively, a digital readout of mean cell volume may be obtained using the MCV/HCT accessory. The Coulter model D industrial costs £1475 and the model ZB1 costs £3200. The Coulter Channelyzer C-1000 and XY recorder cost £4880 (1979 prices).

8.3. Storage of cells

The long-term storage of cells at low temperature is now routine. Methods awaited the observation that glycerol exerts a marked protective effect on the cells during the freezing process. During freezing (particularly during rapid freezing) osmotic changes occur which adversely affect lipoproteins and cause the splitting of cell membranes. Substances which protect cells prevent this excessive concentration of electrolytes. Protective substances must be non-toxic, of low molecular weight and high solubility and must be readily able to permeate living cells (Lovelock and Bishop, 1959). Glycerol is still the most preferred protective agent but some cells (particularly bovine red blood cells) are only permeated slowly. Dimethylsulphoxide (DMSO) is commonly used as an alternative to glycerol.

In general, slow freezing and rapid thawing is recommended for maximum survival. The freezing must not be too slow as this encourages ice-crystal formation, but if too quick there is insufficient time to permit exit of water from the cell. The compromise rate is $1°/min$ which is generally achieved by placing the cells in a box of expanded polystyrene (about 1.5 cm wall thickness) in a $-70°$ C freezer for about 2 h before transfer to liquid nitrogen. At this temperature $(-196°)$ cells may be stored for many years without substantial loss of viability. Some cell lines, e.g. BHK C13 may be stored for many months at $-70°C$ and this is very convenient for those cells in constant use in a laboratory.

8.3.1. Freezing procedure

– Grow the cells for 2–3 days in Roux bottles until a confluent monolayer is formed. It is advisable to change the medium on the cells 24 h before they are due to be harvested.
– Harvest the cells from each Roux bottle into a separate universal container. Count the cells. Take a bacteria check from each bottle and keep 10^6 cells for a PPLO check (Chapter 9).

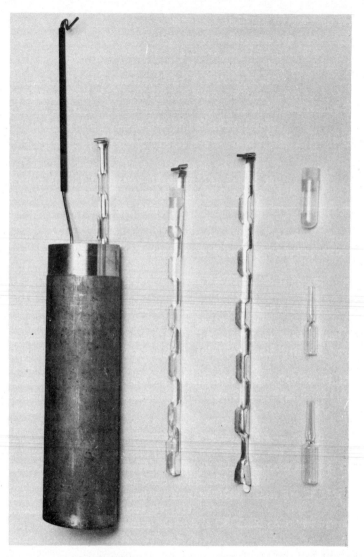

Fig. 8.4. Materials used for freezing cells. On the right are (top to bottom) a plastic ampoule and a 1 ml and a 2 ml glass ampoule. When sealed and cooled to − 70 °C these may be snapped into canes (middle) which are put into the container on the left for immersion in the vat of liquid nitrogen.

- Centrifuge at 250 g for 5 min to pellet the cells. Meanwhile prepare storage medium:
     Eagle's MEM 65 ml
     Calf serum 25 ml
     Glycerol 10 ml
- The glycerol should be sterilised by autoclaving at 15 lb pressure for 15 min
- Resuspend the cells in storage medium to contain 5–6 × 10^6 cells/ml. As an alternative to pelleting the cells the cell suspension may be diluted with glycerol containing medium to give a glycerol concentration of 10%.
- Using a Pasteur pipette, dispense approximately 1.5 ml aliquots into sterile 2 ml blue spot or yellow band, easy snap glass ampoules or sterile plastic ampoules (Fig. 8.4). Glass ampoules are available from Epsom Glass Industries Ltd., and plastic ones from Nunc or Sterilin Ltd. (Appendix 3).
- Seal ampoules carefully (glass ampoules are sealed using the bunsen flame; plastic ampoules are sealed using the screw cap).
- Test for improperly sealed ampoules by immersing the ampoules in a methylene blue/methylated spirit solution. The dye will penetrate those ampoules which are improperly sealed.
- Rinse ampoules in methylated spirit, dry and label.
- Pack well in a polystyrene box and place at − 70°C. The freezing rate should be slow, approximately 1°C/min.
- Transfer to canes (Fig. 8.4) and immerse *CAREFULLY* into the liquid nitrogen vat. Suitable vats with canes are available from Union Carbide or British Oxygen Co. (Appendix 3).
- Note details of cell line in nitrogen stock file. Instead of 10% glycerol, 10% DMSO may be used, in which case the cell suspension must always be held on ice until ready for freezing.
- Gloves and goggles or visors must be worn when handling material in liquid nitrogen.

8.3.2. Recovery of cells from liquid nitrogen

When removing glass ampoules from liquid nitrogen, always observe the following precautions:

    (a) wear a perspex face shield

    (b) wear protective gloves

    (c) leave cane aside for 20–30 s before removing an ampoule. If any ampoules are liable to explode they will do so soon after being taken from the nitrogen.

– Prepare a 100 ml medical flat bottle with 10 ml of growth medium. Gas and warm thoroughly to 37°C.

– Observing all the above precautions remove an ampoule from the cane.

– Thaw the ampoule in a 37°C water bath *as rapidly as possible* using gentle hand agitation.

– Immerse the ampoule in 70% alcohol to sterilise outside.

– Score the neck of a glass ampoule with a file or diamond and break the ampoule between the folds of a sterile towel. Unscrew the cap of plastic ampoules.

– Transfer the contents of the ampoule to the pre-warmed medium in the medical flat bottle. Incubate at 37°C.

– After 24 h decant the medium from the bottle. Keep and test for bacterial contamination. Replace with 10 ml of fresh warm growth medium.

– Incubate at 37°C until cells reach confluence.

8.3.3. Organisation of stocks of frozen cells

Following prolonged subculture of cell strains and cell lines there is a significant chance that the properties of the cells will change. Mutants may arise which will outgrow and replace the original cells. In order to ensure that the cells being used in an experiment remain the same over a period of years, it is therefore essential not to subculture them too often but to return to frozen stocks at least every two months.

On obtaining a cell strain or cell line which is to be used as experimental material over a period of months, one of the first steps is to grow several Roux bottles of cells for freezing. Then every two months a fresh vial should be unfrozen and new cultures set up. As soon as these are available in adequate amounts the old cultures should be discarded. Before the stocks of frozen cells fall too low a new batch of cells should be prepared for freezing. On subculturing cells a note should always be made of the passage number and it should be realised that cells of high passage number may not be identical with the original cells.

As a laboratory may require several canes of frozen vials for each cell type being carried, this puts a certain strain on liquid nitrogen storage facilities. This is, however, the recommended procedure and does provide a reliable back-up in cases of contamination.

8.4. Karyotyping

The chromosome constitution of cells, or cell hybrids, is an important criterion for monitoring the nature of cells in culture. It may help to indicate contamination of the cells of one species with those of another and enables the elucidation of the chromosomal location of various genetic markers. In vitro, however, cells often change their chromosome constitution and most cell lines no longer have the diploid number of chromosomes. Rather they are aneuploid and may have a wide range of chromosomes per cell (see § 2.1).

Although it is easy to see the chromosome mass in metaphase cells, the individual chromosomes cannot be distinguished. In order to analyse the chromosomes present in a cell (karyotype) it is necessary to swell the cells in a hypotonic buffer so that, when they are dried onto a slide, the chromosomes settle down spread out over a wide area. It is important, however, that the cells do not burst in suspension, otherwise it becomes impossible to distinguish the cellular origin of individual chromosomes.

As even in an exponential culture only a small proportion of cells are in mitosis, it is necessary to increase this number by addition of colcemid, which blocks cells in mitosis (§ 10.2). This also has the effect of separating individual chromosomes by its action on the spindle.

8.4.1. Chromosome preparation

The procedure outlined here (Hsu, 1973) is designed to spread out the chromosomes of one cell without leading to their intermingling with those of adjacent cells. (Alternative procedures involve treating a cell monolayer with hypotonic medium before fixation.)

a. Treat exponentially growing cells with colcemid (0.06 μg/ml) for 2 h.

b. Trypsinise monolayers and pellet cells quickly.

c. Suspend the cells in 5 ml hypotonic medium (one part growth medium and 2 parts water) and leave for 10 min.

c'. As an alternative cells may be resuspended in complete growth medium (to stop trypsin action) and then washed twice in Hanks BSS before resuspending in a small volume (0.5 ml) Hanks BSS to which 2 ml of water is then added. (For Q banding 75 mM KCl is recommended – see below.) Allow the suspension to stand in this hypotonic salt solution for 8–10 min.

d. Sediment the cells and add 4 volumes of fixative (3 parts methanol : 1 part glacial acetic acid) without disturbing the pellet.

e. After 10–20 min (depending on the size of the pellet) resuspend the cells and recentrifuge.

f. Resuspend in fresh fixative and recentrifuge.

g. Repeat f twice more, finally resuspending in a small volume (1 ml or less) fixative.

h. Dip a cleaned slide in ice-cold 40% methanol and while still wet add a drop of cell suspension to one end of the slide allowing it to drain downwards.

i. Air-dry and stain with acetic orcein (2% orcein in 45% acetic acid, 3–5 min) or Giemsa (Appendix 2).

j. Rinse, air dry and mount.

A much simpler, though in general less satisfactory procedure, is to grow cells on coverslips and treat as above with colcemid. Treat the coverslip for 30 min with warm, hypotonic saline (see c' above) and then fix for 10 min and dry at room temperature. Stain the cells and mount the coverslips, cells downwards.

8.4.2. Karyotyping

To prepare a karyotype (the chromosome constitution of a cell) a series of metaphase preparations is photographed and the individual chromosomes cut out and arranged in order of decreasing size (an Idiogram – Fig. 8.5). In most cases groups of chromosomes are

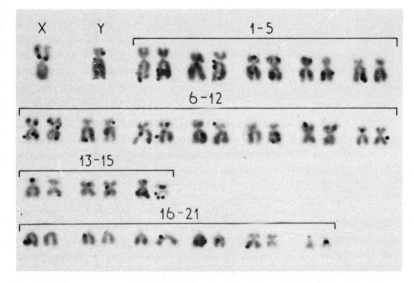

Fig. 8.5. Chromosomes of BHK21/C13 cells. The chromosomes were cut out from a photograph of a metaphase cell chromosome spread and arranged roughly into groups on the basis of size and by reference to the ratios of the lengths of the long and short arms (Marshall, 1972). The autoradiographic grains are the result of an in situ hybridisation with tritiated poly U (Steffensen, 1977). This was achieved by treating the fixed cell preparation with 0.2 N HCl (to remove basic proteins) and ribonuclease (100 μg/ml 2 × SSC) before denaturing the DNA in 0.07 N NaOH for 2 min. The preparation was then washed three times in 70% ethanol and three times in 95% ethanol and dried. 10 μl (140 ng) of [3H]poly U in 3 × SSC was applied to the cover-slip which was incubated at 20 °C overnight before washing thoroughly in 2 × SSC. After a second ribonuclease treatment (20 μg/ml for 30 min at 20 °C) the coverslip was washed in 2 × SSC (three times), 70% and 95% ethanol, air-dried and auto-radiographed using Kodak AR10 stripping film. (Reproduced from Shenkin, 1974, with kind permission.)

apparent – e.g. in man there are seven groups – but seldom is it possible to identify individual chromosomes solely on the basis of size and position of the centromere.

Karyotyping has advanced dramatically since it was discovered that certain stains stained individual chromosomes in characteristic ways. Caspersson (Caspersson et al., 1970a,b, 1971) showed that quinacrine mustard and quinacrine dihydrochloride (Fig. 8.6) pro-

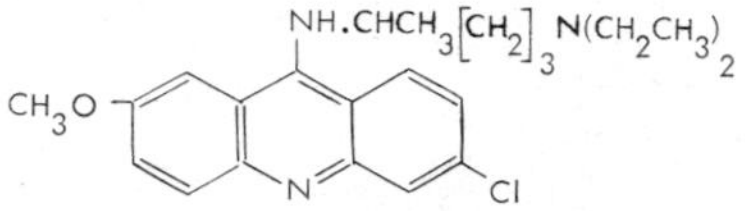

QUINACRINE OR ATEBRIN

Fig. 8.6.

duced a characteristic banding pattern (Q-banding), and later it was found that complementary bands could be formed with Giemsa (G-bands) if the chromosomes are treated first with trypsin or mild alkali (Patil et al., 1971; Seabright, 1971; Wang and Federoff, 1972).

8.4.3. Q-banding (Caspersson et al., 1970a; Lin et al., 1971)

– Use slides on which cells (swollen in 75 mM KCl) have been fixed as described in 8.4.1 a–h.
– Dip the slides into distilled water for a few seconds. Transfer to staining solution (0.5% w/v quinacrine dihydrochloride in 0.1 M phosphate buffer pH 4.5. The stain may be obtained from G.T. Gurr Ltd.; Appendix 3 – under the name 'Atebrin').
– After 15 min wash the slides in three changes of distilled water for a total time of 10 min.
– Air-dry and mount with a drop of distilled water, sealing the edge of the coverslip with paraffin wax or nail polish.
– Examine *as soon as possible* with a fluorescence microscope (excitation filter No. 1 (BG12) and barrier filter No. 47).

The initial Caspersson method uses a staining solution of 5 mg% quinacrine mustard in 0.09 M Na_2HPO_4, 8.7 mM citric acid pH 7.0. After 20 min the cells are washed three times in the same buffer or in distilled water pH 7.0, and sealed wet for visualisation with a fluorescence microscope.

8.4.4. G-banding (Wang and Federoff, 1973; Moorhead et al., 1960)

Treatment of metaphase chromosomes with dilute trypsin solutions brings about a removal or redistribution of non-histone proteins along the chromosomes producing banding patterns observable with the phase contrast or interference microscope (Comings et al., 1973; Stubblefield, 1973). The bands can be intensified by staining with Giemsa when dark bands are apparent corresponding to the bright fluorescent Q-bands.

Growing cells are treated with colcemid (0.4 μg/ml) for 3 h at 37 °C and if they are growing as a monolayer they are then harvested by trypsinisation.

- Pellet the cells (200 g for 5 min) and suspend in 5 ml of hypotonic (0.075 M) KCl.
- Leave at room temperature for 8 min and then add 1 drop of cold fixative (methanol : glacial acetic acid, 3 : 1).
- Mix gently and pellet the cells at 4 °C.
- Remove the supernatant and add fresh fixative gently, without suspending the cells.
- After 30 min resuspend the cells, pellet, resuspend in fixative and pellet the cells again.
- Finally resuspend the cells in 0.25 ml fixative and add a drop to a thoroughly clean (degreased) slide. Air-dry quickly.
- Treat the fixed preparation with trypsin (0.025% versene pH 7.0) for 10–15 min at 30 °C. This causes considerable swelling of the chromosomes. The actual time of treatment should be monitored by phase contrast microscopy – the end point being when the two chromatids fuse together and appear as a unit.
- Rinse in 70%, 80% and absolute ethanol. Air-dry.
- Stain for 12 min with Giemsa (pH 7.0; Appendix 2).
- Wash in distilled water. Air-dry.

Alternatively, the air-dried slide may be transferred to a solution of quinacrine mustard dihydrochloride (50 μg/ml in MacIlvaine's buffer – 6.5 ml 0.2 M Na_2HPO_4, 43.6 ml 0.1 M citric acid and H_2O to 100 ml pH 7.0). After 20 min the slides are washed three times with MacIlvaine's buffer and sealed wet for visualisation with a fluorescence microscope.

8.5   Cell visualisation

A major advantage in the use of cell cultures is that they may be inspected constantly throughout the experiment. As healthy cells are almost transparent this is often done using phase microscopy. However, as the cells are growing on the bottom of a dish or bottle and the depth of the vessel is too great to allow a microscope to focus through it, microscopes had to be developed where the objective was below the microscope stage. Such inverted microscopes (Fig. 8.7) are available from most microscope manufacturers. Two important points to note

Fig. 8.7. Inverted microscopes. From left to right, Wild, Olympus and Prior.

when purchasing an inverted microscope are a) the lamp housing should be adequately ventilated for prolonged use in a 37 °C hot room, and b) the distance between the microscope stage and the condenser should be sufficient to enable the largest roller bottle to be accommodated.

8.5.1. *Phase contrast microscopy* (Gray, 1972; Bradbury, 1976)

This is extremely useful for examining living cell cultures which are almost transparent. Cellular material does change the vibrational state of the light and this may be detected using a phase microscope. The system is based on the phase alteration of light rays that are transmitted through the optical system of the microscope.

1. With the preparation on the microscope roughly in focus adjust the substage condenser to focus the iris diaphragm (partially closed) if necessary centring this to the field.
2. Focus the specimen.
3. Replace one eyepiece with the auxiliary telescope and focus this on the dark ring in the phase plate.
4. Superimpose the bright annulus exactly on the dark phase ring by centring the annulus holder.
5. Replace the normal eyepiece.

Nomarski differential interference contrast microscopy is an alternative to phase contrast microscopy which gives an almost three-dimensional image. For this technique it is essential to use plana-chromat objectives together with a Nomarski interference condenser and this is not available with an inverted microscope.

8.5.2. *Fluorescence microscopy* (Nairn, 1969)

This is commonly used when cell preparations have been treated with a fluorescent antibody (see § 14.3.2) or chromosomes stained with quinacrine (see § 8.4.3) but can also be used to visualise nucleic acids etc., by taking account of their fluorescence when excited by light of wavelength about 260 nm. It is important to use a powerful lamp which produces light of short enough wavelength. Commonly a mercury vapour lamp is used, though a certain amount of work may be done with the cheaper quartz iodide lamp.

The light first passes through a primary filter which only allows through light of wavelength required to excite the fluorescence. When this light passes through the specimen some of it is excited to a longer wavelength. A secondary, barrier filter removes the exciting wavelength allowing only the light of altered wavelength to proceed to the eyepiece.

In the more modern fluorescence microscopes the primary filter has been replaced by a system of mirrors (the Ploem illuminator in the Leitz microscope) which serves a similar function.

CHAPTER 9

Contamination

9.1. Bacterial contamination

All culture media provide an excellent culture medium for the growth of microbial contaminants, and average cell culture procedures (involving the frequent opening of culture vessels) provide ample opportunities for contamination. Moreover, whereas the human body has multiple defences against infection, a bottle of cells relies on the skill of the manipulator. Although antibiotics are commonly added to short-term cultures they are not recommended for long-term cultures. It is essential that stringent precautions are taken to exclude microorganisms and that adequate tests are performed to ensure that the precautions have been effective (see § 9.2).

9.1.1. Glass- and plasticware

The methods of preparation of glassware are indicated in Chapter 3, and if sterilisation is monitored as described the glassware should not be a source of contamination. Likewise plasticware is obtained from the manufacturer in a sterile condition. Usually sterilisation of plastic is achieved using ethylene oxide or irradiation procedures and vessels are supplied wrapped in cellophane.

9.1.2. Cells

Contaminating microorganisms may be present in the cells when these are obtained. The procedures for obtaining primary cells should ensure the exclusion of contamination, but all cells entering the laboratory should be tested before carrying out transfer experiments in rooms being used for other cell transfers.

9.1.3. Media

Certain items, such as balanced salt solutions and versene which are heat-stable, are generally sterilised by autoclaving, but the majority of organic materials used in cell culture are filter-sterilised. For heat-sterilised materials it is generally sufficient to rely on a sterilisation indicator which should be included with each batch of materials being sterilised. Often autoclave tape is sufficient if used on each packet, but for solutions and larger containers, e.g. cans of pipettes, it is recommended that a liquid indicator is included within the bottle or can.

9.2. Sterility checks

Sterility checks should be made on all batches of media and trypsin before these are used. As some tests take several weeks to complete, it is essential that adequate stocks are maintained. The use of untested medium is a sure way to introduce contamination into all cell lines in use in a laboratory. Similarly, cell cultures should be routinely checked for contamination. Such checks will involve growth tests on the medium in which the cells have been growing as well as tests on the cells themselves.

No single test is sufficient to detect all possible contaminants and hence multiple procedures must be adopted. Organisms that grow rapidly in cell culture medium are readily apparent when contaminated medium is incubated for a few days. Such contamination is not a serious problem as the experiment can quickly be terminated and the contaminated culture eliminated. It is the slow-growing contaminant which produces no obvious change in the medium and which exerts no marked cytotoxic effect, which may be overlooked and yet may dramatically interfere with a biochemical investigation, e.g. satellite DNA bands in CsCl density gradient analyses or unusual forms of enzyme may reflect the presence of a contaminant.

In order to detect slow-growing or fastidious contaminants a sample (usually 0.5–1.0 ml) of the culture medium, cell suspension or trypsin solution is inoculated into a series of broths (20 ml) or smeared on nutrient plates. The size of the inoculum may be limiting in the case of very low levels of contamination but is chosen for convenience.

Armstrong (1973) recommends that all freshly prepared media etc. should be tested for possible contamination prior to addition of antibiotics.

- Incubate samples at 37°C for 1 week.
- Remove duplicate 1 ml aliquots into (a) thioglycollate broth, and (b) Saboraud's liquid medium (Appendix 4).
- Also streak samples onto two blood agar plates (Appendix 4) and two deoxycholate plates.
- Incubate as follows:

| | Temperature | | Aerobic | Anaerobic |
	20°C	37°C		
thioglycollate broth	+	+	+	
Sabouraud's medium	+	+	+	
blood agar plate		+	+	+
deoxycholate plate		+	+	+

Various other broths for detecting bacterial and fungal contamination include brain heart infusion broth, tryptose phosphate broth and trypticase soy broth. These should be made up as per manufacturers' (Oxoid Ltd. or Difco Labs.; see Appendix 3) instructions, but some procedures are given in Appendix 4.

Incubation times for the various tests should be at least 7 days, although all but the least obvious contaminations will show after 2 days. The temperature of the incubations should be at both 37°C and at 30°C or 20°C. It is important to maintain the agar plates in a sealed container or humidified incubator to prevent their drying out. Containers are available commercially which not only maintain humidity but also allow anaerobic conditions to be maintained by ensuring the removal of all traces of oxygen. Should a precipitate form in a broth culture, it should be examined microscopically to determine its nature. Such manipulations, however, must be carried out at the end of the working day and must be followed by severe decontamination of the work area and surrounding air space.

9.3. Analysis of bacterial contamination

Using an aseptic technique remove a sample of 'spent' medium and sediment it at 20,000 g for 20 min.

A loop of the sediment should be spread on a blood agar plate (Appendix 4) and incubated at 37°C for 2–7 days when most bacterial contamination will show up.

Samples should be smeared onto glass slides and air-dried prior to staining with 'Gram stain' or methylene blue (Appendix 2).

9.4. Airborne contamination

Despite what has been said above, the major cause of contamination is inadequate aseptic technique allowing organisms from personnel to enter cultures. Such airborne contamination may occur any time a culture vessel is opened.

Consider that if 10 ml of a cell suspension is removed from a vessel it is replaced by 10 ml of air. It is therefore essential to reduce airborne contamination to a minimum. In an undisturbed room bacteria and fungal spores rapidly settle to the floor or the bench, and hence regular cleaning of the floor and bench with antiseptic solutions is required. The floor of the work room should be free of cracks and should be cleaned daily with a disinfectant solution. The work bench should be swabbed down before and after each use with a solution of 70% ethanol. This also serves to kill cultured cells which may have been spilt and hence prevents their transfer to other cultures (see § 2.1).

It is also recommended that the air supply to transfer rooms is filtered through high efficiency particulate air (HEPA) filters which remove virtually all particles larger than 0.3 μm diameter and hence sterilise the air entering the work room.

When an aseptic room is not in use the air may be sterilised by the use of UV germicidal lamps. These should be installed in a position such that the whole room is illuminated and should be sufficiently powerful to be effective at the extreme corners of the room.

Many of these preventive measures are made much easier if, instead of a small room a cabinet is used to carry out aseptic transfer (see below).

In addition to these precautions the necks of all bottles and pipettes etc., should be 'flamed'. This does not have the function of sterilising the glassware but of raising its temperature above the ambient and thus causing an upflow of air around the bottle or pipette. It prevents airborne contaminants from settling into flasks or onto pipettes.

9.4.1. Aseptic technique

The details of a good aseptic technique cannot be written in a book. They must be demonstrated and practiced assiduously under the guidance of a skilled operator. It is, however, largely a matter of common sense, and a few do's and dont's are listed below.

– Always use plugged pipettes and preferably use an automatic pipette or a mouthpiece. This short rubber tube enables you to pipette without bending over the vessels.
– When withdrawing a pipette from a can, do so without touching any other pipettes. This can be done by shaking the can until a few pipettes protrude from the opening. If a second pipette is accidentally touched, it should be removed immediately.
– Replace the lid on pipette cans and bottles as soon as practicable.
– Discard all used glassware which may be a source of contamination into baths of disinfectant (chloros).
– Never leave dregs of medium around to provide a source of bacterial nutrient.
– If you drop a bottle cap – get a new sterile one.

– All bottles of medium prewarmed by standing in a water bath should be thoroughly dried with tissues (not a dirty towel) before being taken into the aseptic transfer room. Water baths should contain bacteriostatic agents, e.g. Panacide (British Drug Houses Ltd.; Appendix 3).

– If in doubt decontaminate the outsides of all bottles with a tissue soaked in 70% ethanol.

If all the above precautions are taken, then the only remaining source of contamination is the person working with the cells. Personnel carry microorganisms on the clothes and on their person, especially on their hair. It is a good idea to wear a clean laboratory coat restricted to use in the cell culture laboratory and never to move from an animal house into a cell culture laboratory without changing laboratory coats. People with long hair should tie it back and wear a clean head covering. It is also recommended that face masks are used especially if there is any doubt about a person's health.

9.4.2. Laminar flow systems

Laminar flow cabinets have a stream of air, filtered through a HEPA filter, entering the cabinet and passing over the work area before leaving the cabinet past the worker. The air flow is piston-like (laminar) and it may be vertical (from ceiling to floor) or more usually horizontal, so that any particulate matter or aerosols generated in the cabinet are cleared within seconds.

A different type of laminar flow cabinet is required when working with cell cultures infected with dangerous viruses. In such cases the aerosols generated within the cabinet may be a biohazard, and outgoing air must not be allowed to enter the general atmosphere until it has been sterilised by passing through a viricidal filter.

The use of a laminar flow system would appear to make it impossible for bacteria to pass from the worker to the culture and in principle such laminar flow systems are the ideal protection of both cultures and workers. However, the very act of putting the hands into the air stream disturbs the laminar flow and causes eddies.

For this reason laminar flow cabinets should only be used in addition to standard aseptic techniques. A laminar flow hood is strongly recommended for those who wish to do cell culture and who cannot set aside the appropriate rooms. They may be used in a general laboratory without fear of contamination if good aseptic technique is also practiced. They have the advantage that the worker is not confined within a small room which may become extremely hot and stuffy, unless ventilated with a very powerful air stream, and secondly that the workers head is separated from the cultures by a perspex screen thus further reducing the chance of contamination.

9.5. Antibiotics

The use of antibiotics in stock cultures is strongly discouraged. The relaxation in aseptic technique resulting from a reliance on antibiotics leads to considerable contamination. The growth of the contaminants may be kept in check by the antibiotics, but biochemical alterations may still be produced. Under the selective conditions antibiotic resistant microorganisms will be selected and once established these may spread like wildfire through all the cultures in a laboratory.

The use of antibiotics should be restricted to short-term experiments. These often involve a large increase in the cell number over a period of a week or so and a transfer of large numbers of cells, and their subsequent experimental manipulation is particularly difficult to perform aseptically. At the end of the experiment such cultures should not be returned to stock but should be treated as contaminated.

9.6. Disposal of contaminated material

No obviously contaminated vessel should be opened. Rather it should be autoclaved immediately. If it is desired to test the nature of the contamination, a sample should be taken aseptically to avoid the spread of infection. This is best done at the end of the working day, and the work area then thoroughly disinfected and the UV lamp left on overnight. Contaminated dishes should be placed in special disposal bags (Sterilin Ltd.) before autoclaving, and often it is advisable to decontaminate the incubator by swabbing with disinfectant.

Contaminated cultures which are also radioactive may be autoclaved (dishes should be placed in disposal bags) and then treated as radioactive waste.

9.7. *Mycoplasmas* (Smith, 1971)

These are small prokaryotic cells (0.3–0.5 μm in diameter) which can form small colonies resembling those of the agent causing bovine pleuropneumonia – hence their alternative name of pleuropneumonia-like organisms or PPLO. The agents do not possess a cell wall and hence will only grow in certain media. When grown under agar, colonies resemble fried eggs in having a thickened central region (Fig. 9.1).

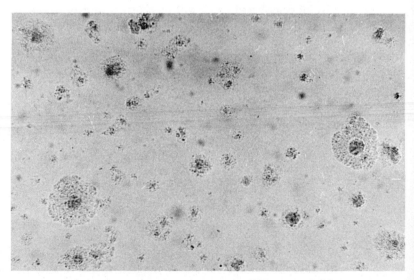

Fig. 9.1. Colonies of *M. arginini* growing on PPLO agar.

There are several serological groups of mycoplasmas from two genera (*Mycoplasma* and *Acholeplasma*). At one time it was considered that they may have been derived from L-forms of bacteria. However, they show no ability to revert to bacterial forms and are now known to represent stable species.

Although primary cell cultures are free from contamination by mycoplasmas, many cell strains and lines in use are contaminated. This probably arises as animal sera are known to be frequently contaminated with *A. laidlawii* and *M. arginini* and because *M. hominis, M. pharyngis* and *M. salivarium* are readily isolated from human mouths and throats.

As mycoplasmal contamination of cell cultures is not always so obvious as bacterial contamination, it is important to 1) be aware of the effects of mycoplasmas on cell cultures, and 2) carry out routine tests for their presence.

9.7.1. *Effect on cell cultures*

The effects will depend to some extent on the species of mycoplasma involved. Thus while *M. gallisepticum* and *M. mycoides* are pathogenic, other species do not produce cell death, but retard the growth of cultured cells. In some cases this effect on growth rate can be attributed to arginine deficiency brought about by high levels of arginine deaminase in, for example, *M. hominis* (Kenny and Pollock, 1963; Kenny, 1973). *A. laidlawii* rapidly cleaves deoxyribonucleosides (Hakala et al., 1963). Thus attempts to incorporate tritiated thymidine into such contaminated cultures are unsuccessful as the nucleoside is rapidly converted to thymine.

9.7.2. *Culture of mycoplasmas*

Mycoplasmas show complex growth requirements (Rodwell, 1969) and hence do not readily grow in simple media such as Eagle's unless growing animal cells are also present. The effect has been attributed in part to the degradation of the DNA of the animal cells by deoxyribonucleases secreted by the mycoplasmas. The liberated nucleosides are essential growth factors for the mycoplasmas (Stock and Gentry, 1971; Stambridge et al., 1971; Kenny and Pollock, 1963).

It is therefore important to test for mycoplasma using media enriched in, for instance, yeast extract which supplies all essential nutrients. In addition, some mycoplasmas grow better under anaerobic conditions.

Kenny (1973) recommends a Soy peptone–yeast dialysate medium (Appendix 4) for culture of mycoplasma. Inclusion of arginine (16 mM) and 0.4 mg% phenol red indicates the presence of arginine deaminase by formation of alkali (purple colouration). Alternatively, incubation with tritiated thymidine and analysis of the culture medium for tritiated thymine can be used to detect thymidine phosphorylase (House and Waddell, 1967).

A simpler medium based on the formula of Hayflick (1965b) is in use in our laboratory.

70 parts 3% brain–heart infusion (Difco Labs.)

20 parts horse serum (preheated to 56°C for 30 min)

10 parts autoclaved yeast dialysate (or extract)

supplemented with 0.5% glucose, arginine HCl and phenol red together with penicillin G (10,000 units/ml) and thallium acetate (0.05%).

1 ml of animal cell suspension is inoculated into 10 ml broth medium.

More often, for diagnostic purposes, however, PPLO checks are carried out using PPLO agar (Appendix 4).

Test for PPLO. Take up a small amount of cell suspension in a Pasteur pipette. Pierce duplicate PPLO agar plates with the pipette about 10 times each. Incubate in 5% CO_2 in either N_2 or air in a sealed jar, i.e. anaerobically or aerobically. Colonies appear in 3–7 days but incubation should be continued for up to 3 weeks when the colonies assume a typical 'fried egg' appearance (Fig. 9.1).

Yeast extract and the various broth and agar media are available commercially from Oxoid Ltd. or Difco Labs. (Appendix 3).

9.8. Staining for mycoplasma

Observation of a monolayer of contaminated animal cells sometimes gives early warning of the presence of mycoplasma. Thus cells stained with haemotoxylin-eosin show the cytopathic effect characterised by multiple dark staining regions (granules).

9.8.1. Orcein stain

Fogh and Fogh (1968) give details of a method using orcein to detect mycoplasma.

- Seed cells onto coverslips so that they are not quite confluent after 48 h.
- Put a coverslip of cells into a Petri dish containing 3 ml 0.6% sodium citrate and slowly add 1 ml of distilled water. Leave for 10 min and then slowly add 4 ml Carnoy's fixative (Appendix 2). Pour off the fluid and replace with 2 ml Carnoy's fixative.
- Allow 10 min for fixation. The coverslip is then allowed to stand for 5 min or until absolutely dry. Stain for 10 min with orcein stain (Appendix 2) or longer if necessary and wash 3 times in absolute alcohol.
- All the procedures are carried out at room temperature.
- Examine using the phase microscope – infected cells show mycoplasma located primarily at the cell borders and apparently in part attached to the peripheral parts of the cytoplasm and in the intracellular spaces.

9.8.2. Autoradiography

Autoradiography of contaminated cultures labelled with tritiated thymidine shows grains over the whole cell rather than localised to the nucleus. This is a result of degradation of the thymidine to thymine followed by incorporation into mycoplasma DNA and other cell constituents.

- Seed cells at 2.5×10^4 per 35 mm plate containing a coverslip and label for 30–46 h with 5 μCi [^{3}H]thymidine per plate.

- Fix cells with ethanol:acetic acid (3:1) for 10 min in the dish. Repeat and wash twice with absolute ethanol.
- Mount the coverslip on a glass slide – cells uppermost – and dip into Ilford L4 emulsion (see § 12.4.4).
- After 5–7 days develop in Kodak D19b developer and fix.
- Examine under the microscope. Cells free of PPLO show heavy nuclear labelling, whereas infected cells show cytoplasmic labelling with little or no nuclear labelling.

9.8.3. Fluorescence staining

The presence of mycoplasmal DNA in the cell cytoplasm may be detected by staining with the fluorochrome Hoechst 33258 (Appendix 3). This intercalating dye fluoresces under ultraviolet light and this forms the basis of a very rapid sensitive test for mycoplasmas.

A stock of the Hoechst 33258 bisbenzamid fluorochrome solution is made by dissolving 5 mg in 100 ml PBS-A (Appendix 1) using a magnetic stirrer. It may be sterilised by filtration through a 0.22 μm membrane and should be stored in the dark at 4°C. It should be diluted a thousand-fold with PBS-A for use.

- Set up coverslip cultures of the cells to be tested and use when 50–70% confluent.
- Aspirate the medium from the cells and fix with two changes of ethanol:acetic acid (3:1).
- Wash in deionised water and incubate for 30 min at 37°C with diluted bisbenzamid fluorochrome.
- Rinse with deionised water and mount the coverslips – cells down – on a slide using glycerol mountant (22.2 ml 2.1% citric acid; 1 H_2O; 27.8 ml 2.8% disodium hydrogen phosphate; 50 ml glycerol pH 5.5).
- Examine with a fluorescence microscope when the presence of mycoplasma shows up as a cytoplasmic fluorescence (Fig. 9.2).

Russel et al. (1975) report a similar staining method using 4',6-diamidino-2-phenylindole (DAPI):
- Rinse cells with PBS.

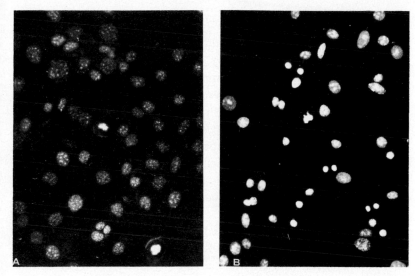

Fig. 9.2. Fluorescence staining for mycoplasma. Cells were stained with Hoechst 33258 which fluoresces with DNA. On the left are 3T6 cells which show positive cytoplasmic staining, particularly obvious in the dividing cells. On the right are negative PyY cells. (The positive 3T6 cells were kindly provided by Dr. D. Lewis of Flow Laboratories.)

– Stain with DAPI at 0.1 μg/ml at 37°C. Although staining is faint after 15 min, it is easier to distinguish cytoplasmic fluorescence with shorter incubation times than after 60 min when nuclear fluorescence is intense.

– Rinse with PBS and mount the coverslip as described above.

This method, using unfixed preparations, is even more rapid than the one described above using Hoechst 33258. A kit based on DAPI is available from Bioassay Systems.

Autoradiography and staining methods are based on detection of cytoplasmic DNA (in the former case replicating cytoplasmic DNA). For this reason there will always be a low background caused by mitochondrial DNA and samples to be tested should be compared with positive and negative controls.

9.9. Elimination of mycoplasmas

Most contaminated cultures should be autoclaved to prevent spread of the mycoplasmas. Occasionally, a very important cell strain may become or be found to be contaminated, and it has been reported that kanamycin (0.1 mg/ml) prevents the growth of some mycoplasmas while having no effect on cell growth (Fogh and Hacker, 1960). The antibiotic tylocine (6.0 μg/ml) is also reported to be effective against PPLO (Gibco Biocult catalogue; Appendix 3).

However, few generalisations can be made; Birskirk (1967) has used a cocktail of antibiotics and such treatment at elevated temperatures (41°C) has been reported to be doubly effective (Johnson and Orlando, 1967).

Mycoplasmas can be eliminated from cell culture by treatment with immune serum (Pollock and Kenny, 1963) and passage through an animal is often effective in removing mycoplasmas from tumour producing cell lines.

9.10. Viral contamination

Since the possibility was recognised that bovine serum may be contaminated with several viruses (e.g. bovine herpesvirus and parainfluenza 3-virus) sera are now routinely tested by the manufacturer. The effectiveness of this screening has, however, been questioned (Kniazeff, 1973).

It is also possible to introduce viruses into a culture through the use of contaminated trypsin, but a far greater problem is the presence of viruses in the cells themselves – particularly primary cells. As many animal species have latent viral infections which cause no overt symptoms, many primary cell cultures carry these viruses. The problem is accentuated by the fact that many cells are taken from tumours which have resulted from a viral transformation event and which may therefore carry all or part of the viral genome in one form or another. The act of preparation of a primary culture may lead to the induction of virus production and an accompanying cytopathic effect. The problem is more serious, however, if no cytopathic effect is seen but the virus continues to replicate at a slow rate or in a few cells only. Thus primary cells from children with Burkitt's lymphoma show no signs of Epstein–Barr virus (EBV) until several passages when a few cells show positive immuno-fluorescence (Henle and Henle, 1966).

Cells transformed with the papovaviruses polyoma (e.g. BHK 21PyY, Dulbecco, 1968) and SV40 (e.g. SV28, Wiblin and Macpher-son, 1972) show the presence of tumour antigen by immuno-fluorescence, but this is not harmful; it rather results in an increased growth potential and consequent disregard for neighbouring cells.

Temperature-sensitive variants of the BHK21 PyY cells have been selected, which at the non-permissive temperature are unable to make T-antigen, and the host cell reverts to its non-transformed character (Dulbecco, 1969). Thus the expression of viral functions depends on the conditions of culture, and often virulent virus may be rescued from non-permissive cells by fusion with uninfected permissive cells (Green, 1970). A similar in vitro rescue can be used to detect the presence of cryptic virus in non-permissive cells. Thus a homogenate of the suspect cells is added to a permissive cell line or injected into a susceptible animal when the presence of a cytopathic effect indicates a positive response (Vigier, 1970).

CHAPTER 10

The cell cycle

10.1. Description

Growing cells are characterised by a sequence or sequences of events leading to duplication of their constituents. These events appear to occur in a strict, temporal order, and growing cells may be considered as a simple system for the study of gene expression. The two most obvious events which occur in growing cells are cell division and DNA synthesis which are the markers used to characterise the cell cycle (Fig. 10.1).

M or mitosis is the period when the cells divide and S is the period of DNA synthesis while G1 and G2 represent gaps – gaps in our knowledge of obvious markers in these areas. Much cell biology in recent years has been devoted to attempts to fill these gaps. These symbols (G1, S, G2 and M) are used throughout the text to describe the phases of the cell cycle.

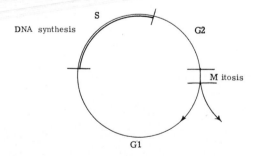

Fig. 10.1. The cell cycle. S is DNA synthetic period. G1 and G2 are the gaps between Mitosis (M) and S, and S and M, respectively.

10.2. Mitosis

The stages of mitosis are:
  prophase
  metaphase
  anaphase
  telophase

Mitosis is heralded by the rounding up of the cell, and the first visible indication that it is about to divide is a change in appearance of the nucleus. This is caused by the condensation of the chromosomes in early prophase, a process which continues as the nuclear membrane disappears, so that by metaphase the highly condensed chromosomes are massed in the centre of the rounded-up cell. At this stage the cells are only loosely attached to the substratum and are readily dislodged by agitation or by gentle trypsinisation (Fig. 10.2). This forms the basis of the method of selection synchrony of mitotic cells (see § 11.2).

The centrioles migrate to opposite poles of the cell and the mitotic spindle is formed, apparently joining the cell membrane through the centrioles to the centromere of each chromosome. Spindle fibres consist of one type of protein, tubulin, of molecular weight 60,000. It is the organisation of these molecules to form the mitotic spindle which is blocked by the drugs colchicine, colcemid, vincristine and vinblastine (Fig. 10.3) with the consequence that mitosis is arrested in metaphase.

At anaphase the two sets of chromosomes move to opposite poles due to tubulin action, and by telophase the chromosomes decondense as the cell membrane encloses each new daughter cell. The paired cells are still rounded up and resemble a dumbbell, but very soon they flatten out as the nuclear membrane and nucleoli reform and the cells enter G1.

10.3. S-phase

S-phase is, by definition, the period in the cell cycle during which DNA is synthesised. It is clear, however, that DNA synthesis does

Fig. 10.2. Mitotic stages. Mouse L929 cells, pulse labelled with tritiated thymidine, were processed for autoradiography. Among the S-phase cells (covered with grains) are cells in other stages of interphase and several mitotic cells. A metaphase cell, two anaphase cells and a late telophase cell are distinguishable.

colchicine

colcemid has CH_3 in place of $COCH_3$

Vincristine

Vinblastine

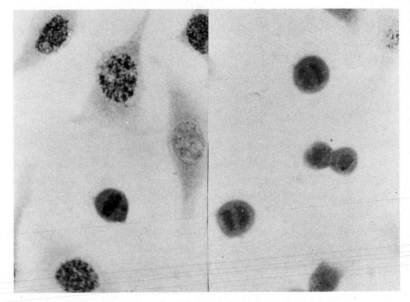

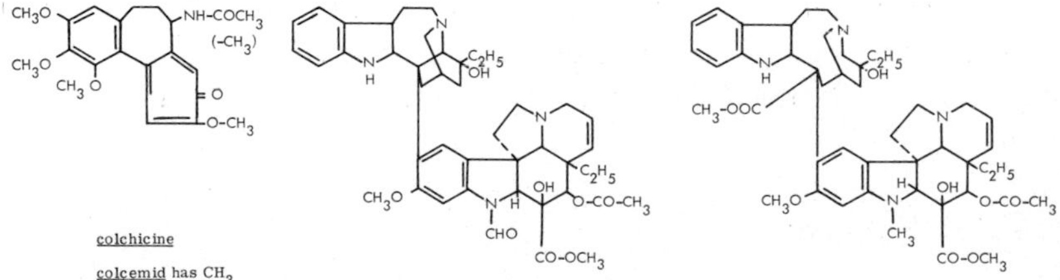

Fig. 10.3. Mitotic blocking agents.

not start suddenly, proceed at full speed, and then stop suddenly, and hence to define exactly the beginning and end of S-phase is virtually impossible.

It has been shown using DNA fibre autoradiography (Cairns, 1966, 1972) that in animal cells the synthesis of DNA occurs in short stretches or *replicons* (Huberman and Riggs, 1968). Replicons vary in size from 15 to 60 μm but are mostly less than 30 μm long in tissue culture cells (but see Callan, 1972). DNA synthesis is initiated in the middle of a replicon and proceeds bidirectionally to the ends of the replicon (Hand and Tamm, 1974). Later adjacent replicons fuse and the two replicated chromatids can then separate (Kowalski and Cheevers, 1976).

At the beginning of S-phase a few particular replicons initiate DNA synthesis. That it is always the same replicons that are replicated at the beginning of S-phase can be shown by synchronising cells at the G1/S boundary using an aminopterin blockade (§ 11.8). Cells are then released from inhibition in the presence of a density label such as bromodeoxyuridine (Kajiwara and Mueller, 1964). The heavy analogue is replaced in the medium by thymidine after 15 min, and the cells are grown for several generations. When the cells are resynchronised and released in the presence of tritiated thymidine, radioactivity is first found in DNA of hybrid density, i.e. bromodeoxyuridine substituted DNA is replicated at the beginning of S-phase.

At the other end of S-phase one of the X-chromosomes in cells of female mammals has been found to replicate later than any other DNA in the cell (e.g. Gilbert et al., 1965). And in between it is clear that particular satellites replicate at particular times in S-phase. Thus there is a specific temporal order in which particular replicons replicate.

This was shown more clearly by Stubblefield and Mueller (1962), who demonstrated focalised synthesis of DNA by pulse-labelling a random population of cells with tritiated thymidine and then after varied time intervals visualised autoradiographically the regions of metaphase chromosomes where incorporation had occurred.

If cells are synchronised at the G1/S boundary and then released, the rate of DNA synthesis is initially slow but accelerates to reach a maximum at about 3 h and then decelerates until S-phase is essentially complete in 6–7 h (Stubblefield and Mueller, 1962; Adams, 1969b). As replication occurs different numbers of replicons are active at any one time, and so it is not surprising that more careful labelling reveals bursts of tritiated thymidine incorporation throughout S-phase rather than a steady even progression (Klevecz, 1969; Lett and Sun, 1970; Klevecz et al., 1974).

Detailed analysis using several different methods (pulse and continuous label with tritiated thymidine, DNA fluorescence per cell using the fluorescent Feulgen assay and flow microfluorography (Van Dilla et al., 1969), and DNA specific fluorescence using the diaminobenzoic acid assay (Kissane and Robins, 1958)) reveals that the DNA content of a cell increases in a saltatory fashion, and that the early portion of S-phase is a period of low net DNA synthesis which may be mistaken for G1 if insensitive methods of measurement are used (Klevecz et al., 1975).

10.4. Control of the cell cycle

Although cells in culture and cells still in the body have similar durations for S, G2 and M, it is apparent that variations in the duration of G1 can be dramatic and in fact account for the major variation in cell cycle time of different cell types (Table 10.1). Thus G1 is apparently absent in cultured Chinese hamster lung cells (Robbins and Scharff, 1967) and no upper limit has yet been placed on its duration. When Chinese hamster cells lacking a G1-phase are fused with $G1^+$ cells (§ 13.5) the $G1^-$ state is dominant (Liskay and Prescott, 1978).

Cells which have spent a long time in G1 lose some of the enzymes typically present in dividing cells – particularly those concerned with DNA synthesis. These cells have traditionally been said to be 'out of cycle' or in G0. The implication of this G0 phase is that to leave G0 cells require a stimulus to urge them past a barrier and back into

TABLE 10.1

Duration (h) of cell cycle phases in cultured cells

Cell type	T	$tG1$	tS	$tG2$
HeLa	20–28	8–16	5–9	2–8
Human fibroblasts	16–30	3–16	6–11	4–5
Human amnion	19.4	9.8	6.7	2.2
Mouse L	18–23	6–11	6–12	3–4
Mouse L5178Y	11.5	1.5	7.1	2.9
Chinese hamster fibroblasts	12–15	3–6	4–8	2–3

Results have been pooled from several sources (Firket, 1965; Cleaver, 1967; Lipkin, 1971; Puck, 1972) to give a general range of data which should be compared to the results for human amnion cells (Sisken and Morasca, 1965) and mouse L5178Y cells (Defendi and Manson, 1963). As the duration of mitosis is short and not always reported, where known it has been shared between $tG1$ and $tG2$. T is the total cell generation time; $tG1$, tS and $tG2$ are the durations of the G1, S and G2 phases respectively. (See Fig. 10.1 for a diagrammatic representation of the cell cycle.)

cycle. Pardee (1974) has suggested that whenever cells are exposed to suboptimal physiological conditions they enter a quiescent phase, and that there is a single restriction point in G1 which regulates their re-entry into a new round of the cell cycle.

There are many experimental conditions where cells are put into or taken out of G0 in order to study the changes associated with onset of proliferation. The following examples are referred to elsewhere in this book:

1. Lymphocytes may be stimulated to grow by mitogens (§ 6.3).
2. Serum starved cells may be stimulated by readdition of serum (Burk, 1970, and § 11.6).
3. Cells which have ceased to grow exponentially and reached a plateau phase because of limitations of growth surface may be stimulated by subculturing (Stoker, 1972, and § 11.5).

However, even in these non-growing cell populations a small number of cells are found to be making DNA or dividing, and it has been suggested that although most of the cells are in G0, a small proportion has been stimulated into proliferation. The question is: does the chance of a cell leaving G0 and entering into cycle depend on the length of time, since it last divided, or is there an equal probability for any cell to enter the cycle? This latter alternative has been expressed most forcibly by Smith and Martin (1973, 1974). They have named S-phase, G2, M and part of G1 the B-phase, and suggest that the duration of the B-phase is fixed within narrow limits for a particular cell type. Shortly after mitosis cells enter the A-state in which the cells do not progress towards division. A cell may remain in the A-state for any length of time but always has a fixed probability (P) of leaving for the B-phase, provided environmental conditions remain constant. Smith and Martin suggest that modification of the transition probability provides a major means of controlling cell proliferation.

The two models of cell proliferation differ largely in the prediction of the Smith and Martin model that the proportion of cells remaining in interphase (α) should decline exponentially with age, beginning at time T_B after mitosis. Thus a semilog plot of α against age after mitosis should show a linear decay after a lag period equal to T_B. As shown in Fig. 10.4 this is borne out by the data. In fact there is an initial downward curvature before linearity is attained caused by the variability in T_B.

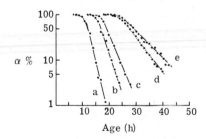

Fig. 10.4. Distribution of generation times of various cell types in culture. The proportion (α) of cells remaining in interphase at various times after division has been recorded for various cells by time lapse cinematography. a) Rat sarcoma, b) HeLa S3, c) mouse fibroblasts, d) L5 cells, and e) HeLa. (Reproduced from Smith and Martin, 1973, with kind permission of the authors and publishers.)

The alternative model would predict that the greater the age of a cell the greater is its probability of division and hence the semilog plot of α against age would constantly curve downwards. In practice it is exceedingly difficult to distinguish the two possibilities and there is considerable discussion as to which model is correct (Baserga, 1978).

Treatment of serum deprived (§ 11.6) 3T3 cells with serum for increasing lengths of time results in increasing numbers of cells being stimulated to enter S-phase. Brooks (1976) argues that serum determines the rate of commitment of cells by maintaining an environment conducive to a high transition probability. The 14-h lag before cells enter S-phase is required to attain an increase in probability. On removal of serum the probability of a cell making the transition to the B-phase drops and the number of cells entering S-phase falls abruptly within 3–5 h. This shows that the commitment step must be less than 5 h before the beginning of S-phase.

Smith and Martin (1974) added 3 mM hydroxyurea (an inhibitor of DNA synthesis; § 11.8.4) just prior to the time serum treated cells were due to enter S-phase. On removal of the inhibitor 6 h later cells entered S-phase, but not all together; rather they entered at a rate dependent on the serum concentration. This was interpreted as showing a continued dependence on the transition probability and was taken to show that the cells were held in A-state in the presence of hydroxyurea and that the commitment step is at the G1/S phase boundary.

The traditional model of the cell cycle explains differences in generation times between the cells in a population as a result of the sum of small differences in the times taken for the large numbers of steps required for a cell to progress from one division to the next. In the model of Smith and Martin (1973), although such differences do occur and will lead to the heterogeneity in the length of the B-phase, the major contribution to the variation in generation time between cells is the length of time spent in the A-state. Minor and Smith (1974) argue that variations in the duration of the B-phase for sibling pairs should be minimal, and they find a strong correlation between intermitotic times of siblings and the overall variability of the population.

Small and large cells from a quiescent culture have been sorted (using the fluorescence activated cell sorter; § 10.7.5) and their cycle times measured (Shields et al., 1978). Both sizes of cell show the same transition probability for leaving the A-state, but the smaller cells spend longer in the G1 part of the B-phase than do the larger cells. In this respect the results are more easily interpreted in the Smith and Martin (1973) model than on the more traditional model with a G0-phase.

10.5. Distribution of cells around the cycle

For every cell which enters mitosis at the end of the cell cycle two will begin the next cycle. This means that the distribution of cells around the cycle is not uniform, but that there is a preponderance of young cells during exponential growth.

In the ideal situation all cells in a population have the same generation time given by the distribution function

$$f(t) = kn_0 e^{-kt}$$

where k = growth constant, n_0 = number of cells at zero time, t = time expressed as a fraction of the generation time, i.e. varies from 0 to 1.

This can be expressed to the base 2 when

$$f(t) = n_0 2^{-t}$$

Thus the ideal distribution of cells around the cycle is shown by the solid line in Figure 10.5 (Engleberg, 1961; Kubitschek, 1966).

In practical terms because of the variability in generation times the dotted line in Figure 10.5 more accurately reflects the real situation (Sisken and Morasca, 1965).

If the cells do not all have the same generation time then the time

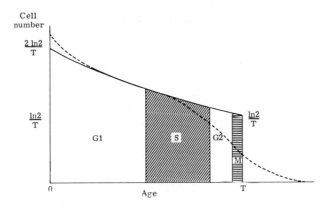

Fig. 10.5. Distribution of exponentially growing cells around the cell cycle. Cells in an exponentially growing population are theoretically distributed round the cell cycle as shown by the solid line. However, as the cell cycle time varies amongst the cells the dotted line more closely resembles the observed distribution. Cells which have just divided have an age of zero while those in the next mitosis have age T. The positions of G1, S and G2 are shown for a typical cell. (Reproduced from Cleaver, 1967, with kind permission of the author.)

taken to double the number of cells (T_D: the cell doubling time) will be slightly shorter than the generation time (T). Alternatively, if not all the cells in the population are growing, i.e. the growth fraction is less than 1 (see below) then T_D will be longer than T.

As well as there being twice as many cells at the beginning of G1 as at the end of G2, similar but smaller differences will exist between cells at the beginning and end of S-phase. Moreover, the proportion of cells in S-phase not only depends on the relative lengths of S-phase and the generation time but also on the location of S-phase within the cycle (but see 'growth fraction' below). It is, therefore, not only inaccurate to say that, if 30% of cells are in S-phase (labelling index LI = 0.3) then the duration of S-phase (tS) is 30% of the generation time (T), but it is even inaccurate to say that

$$LI = \ln\frac{2^{tS}}{T}$$

Rather, the location of S within the cycle must be defined and this is most easily done by relating it to T and $tG2$ to give

$$LI = [\exp tS - \ln\frac{2}{T} - 1] \exp tG_2\ln\frac{2}{T}$$

(Cleaver, 1965).

Because the location of mitosis is fixed and its duration short, one can, however, say that the proportion of cells in mitosis, the mitotic index,

$$MI = \ln\frac{2^{tM}}{T}$$

(Smith and Dendy, 1962).

However, a generalisation is that, while G1 cells have a single complement of DNA and G2 cells have twice this amount, exponentially growing cells have an average of about 1.3 times as much DNA as does a G1 cell.

10.6. Growth fraction

In most cell populations not all the cells are proliferating, there being a proportion of non-proliferating cells. These may alternatively be described as being in G0 or in the A-state. The proliferation index or growth fraction is given by

$$\frac{Nc}{N} = \frac{\text{number of proliferating cells}}{\text{total number of cells}}$$

The time taken for the total number of cells to double (cell doubling time) is therefore not equal to the cell cycle or cell generation time. It is important, therefore, when doing cell cycle analyses to carefully distinguish between T and T_D and preferably maintain the growth fraction as high as possible.

Low growth fractions are common in vivo (even some tumours may have a growth fraction less than 0.1), and very often in vitro in primary cultures or under suboptimal conditions not all the cells are growing. The low growth fraction may arise

by irreversible differentiation
by cells entering G0 or A-state
by cell death.

In the steady state, additions to the population are balanced by subtractions from it and the age distribution is rectangular (Fig. 10.6),

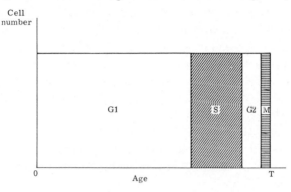

Fig. 10.6. Distribution of steady state cells around the cell cycle. In the steady state the number of cells gained at cell division is exactly balanced by the number lost by death or migration and hence the number of cells in a particular phase is proportional to the duration of that phase.

i.e. the fraction of cells in a phase is now directly proportional to the duration of that phase, e.g.

$$\text{mitotic index (MI)} = \frac{t\text{M}}{T}$$

Further information on cell cycle kinetics can be found in Cleaver (1967) and in Aherne et al. (1977).

10.7. Cell cycle analysis

10.7.1. Tritiated thymidine pulse method (Howard and Pelc, 1953)

This is perhaps the earliest method used, and the following description is for cells growing in small dishes or on coverslips. The method, however, can be modified for use with suspension cultures.

- Set up a series of coverslips in multiwell dishes (see § 2.2). 48 or more coverslips are required, i.e. 2 multiwell trays. Each well should be seeded with $1–2 \times 10^4$ cells in 0.5 ml medium and the cells allowed to incubate overnight to ensure exponential growth. It is preferable to use Hepes buffered medium as the regular harvesting required makes it very difficult to maintain pH with a bicarbonate buffer. It is also important not to let the temperature fall during sampling which should be done in a 37°C room.
- Pulse label each culture for 30 min with [^{3}H]thymidine (5 μCi/ml; 2 Ci/mmol), i.e. to each well containing 0.5 ml growth medium add 10 μl of a solution containing 2.5 μCi [^{3}H]thymidine at a concentration of 1.25×10^4 M.
- Remove the [^{3}H]thymidine containing medium using an unplugged Pasteur pipette connected to a vacuum pump through a trap. Quickly replace with prewarmed medium containing 2×10^{-6} M thymidine.
- Fix coverslips at frequent intervals (20–30 min) by dipping the coverslip into PBS (twice); 5% cold trichloroacetic acid (4 times) and absolute ethanol (twice). Attach the coverslips to one end of a glass slide (cells uppermost) using DePeX (Pearse, 1953) and process for autoradiography (see § 12.4). DePeX is a mountant available from G.T. Gurr Ltd. (Appendix 3).
- Record the proportion of mitotic cells labelled. In theory the first labelled mitotic cell will appear after a time equal to the length of G2 (tG2) and the percentage of labelled mitotic cells should rise to 100 over tM (Fig. 10.7). After a further tS the percentage of labelled mitotic figures should fall and will only rise again after tG1 + tG2.

In practice, because of the variability between cells and difficulties in producing short pulses all boundaries are blurred. However, by taking 50% values for the various transitions a reasonable approximation to the duration of the cell cycle phases can be obtained.

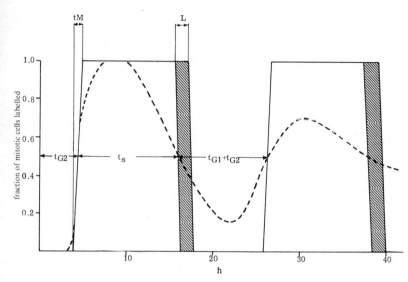

Fig. 10.7. Fraction of labelled mitoses. The solid line represents the theoretical change in the fraction of mitotic cells after labelling for 100 min with tritiated thymidine. The dotted line follows data obtained with mouse L cells. The hatched areas represent the duration of the exposure to tritiated thymidine. (Reproduced from Cleaver, 1967, with kind permission of the author.)

10.7.2. Continuous labelling method

– The cells should be set up and labelled as before except that [³H]thymidine is used at 2.5 μCi/ml (0.06 Ci/mmol).
  Do not remove the growth medium with the tritium labelled thymidine but continue to incubate and harvest coverslips.
– Process for autoradiography and record the fraction of labelled mitoses, the fraction of labelled cells, and the grain count, i.e. the average number of autoradiographic grains over the labelled cells (Fig. 10.8).

The fraction of cells labelled will reach 1.0 only when all the cells not initially in S-phase have proceeded around the cycle and entered S-phase, i.e. after $T-tS$ which in this case = 16.3 h = $tG1 + tG2 + tM$. $tG2$ can be found as before by the time taken for the first

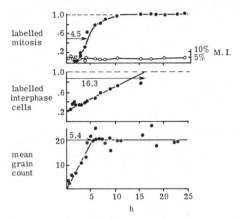

Fig. 10.8. Continuous labelling method. Human skin epithelial cell cultures were exposed to tritiated thymidine (2.5 μCi/ml; 0.06 Ci/mol) continuously for 25 h during which time cells were harvested and scored for the indicated parameters. (Reproduced from Cleaver, 1967, with kind permission of the author.)

labelled mitosis to appear. Thus tG1 can be found, as the mitotic index (proportion of mitotic cells):

$$MI = t_n ln\frac{2}{T}$$

(Smith and Dendy, 1962).

The grain count will plateau when the labelled cells have progressed through the whole of S-phase in the presence of [³H]dThd. However, when these cells divide the grains are now shared between the two cells and the grain count will fall. This introduces errors into the estimated tS. Errors also are introduced by the presence of those cells already in S-phase when the thymidine is added or when the cells are harvested. This method therefore overestimates the duration of tS.

10.7.3. Accumulation functions

A certain amount of information can be gained by measuring the accumulation of mitotic cells on addition of a mitotic blocking agent.

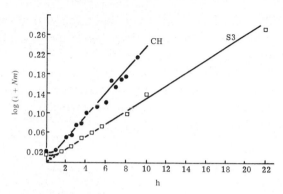

Fig. 10.9. Accumulation of mitotic cells. The mitotic collection function (log $(1 + Nm)$) is plotted against time for Chinese hamster cells $(T = 12.4$ h) and HeLa S3 cells $(T = 20.1$ h) to which colcemid was added at zero time. (Reproduced from Puck, 1964, with kind permission of the author and publisher.)

Cultures may be set up as before and colcemid (0.25 μg/ml) added at zero time.

Figure 10.9 shows that after a short lag the plot of the accumulation function (log $(1 + Nm)$ where Nm = fraction of mitotic cells) against time is linear. From such a curve the generation time (T) can be calculated either from the equation

$$\log (1 + Nm) = 0.301 \, t\text{M} + t/T$$

where tM is the time spent in mitosis and t is the time from addition of colcemid, or by measuring the time required for the accumulation function to reach 0.3. In practice this seldom happens in the theoretical manner as (1) the generation time differs from cell to cell, (2) 100% of the cells may not be viable, (3) some cells may escape the colcemid block and re-enter G1, and (4) colcemid may have deleterious effects on other parts of the cell cycle thus affecting the generation time. The lag in the rate of accumulation of mitotic cells immediately following addition of colcemid is explained by the fact that cells already in mitosis on addition of the drug are not blocked but complete mitosis and re-enter G1. Taylor (1965) has shown that with the related drug colchicine this happens at concentrations between 50 and 200 nM, but at concentrations above 200 nM all cells are blocked in mitosis.

If tritiated thymidine is added along with colcemid three accumulation functions can be measured: (a) total labelled cells, (b) total mitoses, (c) labelled mitoses, which enables calculation of the length of S and G2 and hence also G1.

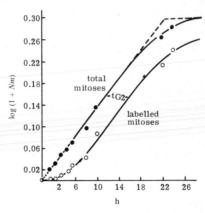

Fig. 10.10. Accumulation of labelled mitotic cells. A similar experiment with HeLa S3 cells to that shown in Fig. 10.9 except that [^{3}H]thymidine was added with the colcemid at zero time. After an initial lag the lines for accumulation of all mitotic cells and labelled mitotic cells are parallel and separated by tG2. (Reproduced from Puck and Steffen, 1963, with kind permission of the authors and publisher.)

Thus from Figure 10.10 it can be seen that an accurate measure of tG2 can be obtained from the distance between the lines which should be parallel if the cell population is exponential and if the tritiated thymidine is not having any deleterious effects on growth. The generation time (T) can be obtained from the slope of the line which equals $0.3/T$.

Using the figure for tG2 so obtained a second graph may be drawn (Fig. 10.11) of time against $\log\left(1 + \dfrac{N(\mathrm{L})}{k}\right)$ where $N(\mathrm{L})$ is the

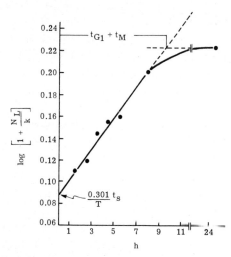

Fig. 10.11. Accumulation of labelled cells. The data points for this experiment can be obtained from the experiment described in Fig. 10.10. The collection function is $\log\left[1 + \dfrac{N(L)}{k}\right]$ where $N(L)$ is the total number of labelled cells and $k = 2^{tG2}/T$. (Reproduced from Puck and Steffen, 1963, with kind permission of the authors and publisher.)

fraction of cells labelled and $k = \exp\ (ln\ 2^{tG2}/T)$ (Puck and Steffen, 1963; Puck et al., 1964). This function reaches a plateau when all the cells in G1 have entered S-phase (or, if an intermediate concentration of colchicine is used, all the M + G1 cells have entered S). As there is significant variation in the lengths of G1 among the cell population this is not a sharp cut off, but the line obtained from the points over the first 6 h or so can be extended, and all that is required is a final figure which can be obtained at 24 h. The intercept 0.3 tS/T gives a value for tS.

Although this method involves a more difficult mathematical background (for which the original references or Cleaver (1967) should be consulted) it involves fewer time points spread over a shorter period.

10.7.4. Graphical analysis

Perhaps the simplest method of cell cycle analysis is that described by Okada (1967) which gives the additional information of the proportion of cells in each phase of the cycle.

The information required from an exponentially growing culture of cells is:

    a. the generation time which may be obtained as the cell doubling time or as an accumulation function as described above (see § 10.7.3)

    b. the mitotic index (see § 10.5)

    c. the percent of cells in S-phase obtained by pulse labelling with [³H]thymidine (see § 10.7.1)

    d. the duration of G2 obtained either from the time after addition of [³H]thymidine to the appearance of a labelled mitotic cell or from the separation between the lines in Figure 10.10.

All this information may be obtained in an experiment lasting 3–4 h with or without the use of colcemid. The graph (Fig. 10.12) is then plotted on single cycle semi-log paper as follows.

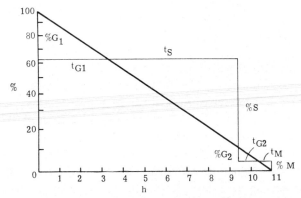

Fig. 10.12. Graphical analysis of cell cycle. 5 cm Petri dish cultures of *Aedes albopictus* mosquito cells were set up and the number of cells per dish counted regularly to establish the doubling time (taken as the generation time). When it was clear that the cells were growing exponentially [³H]thymidine and colcemid were added and cells processed to establish % cells labelled, mitotic index and *t*G2. The graph was drawn as described in the text (see § 10.7.4).

- Mark off the generation time on the linear axis and divide the region from 1 to 2 on the log axis into 100 divisions (percentages), putting 0% at 1 and 100% at 2. Draw a straight line joining 100% to T.
- From T draw a vertical line a length corresponding to the percentage of cells in mitosis and join this to the diagonal line to give the duration of M.
- Continue the horizontal line for $tG2$ and join this to the diagonal line to give the percentage of cells in G2.
- Draw a vertical line corresponding to the percentage of cells in S and join this to the diagonal to give tS.

The final horizontal and vertical lines give $tG1$ and the percentage of cells in G1.

10.7.5. Flow microfluorometry

An instrument developed at the Los Alamos Scientific Laboratory permits analysis of large numbers of single cells with respect to DNA and protein content, cell volume, etc. (Fulwyler, 1965; Kraemer et al., 1973; Klevecz et al., 1975; Herzenberg et al., 1976). The output data from a large population of cells consists of, for example, a distribution of the values of cellular DNA content or cell volume. In addition to the capability for complex analyses the instrument is able to physically separate particular cell subpopulations of interest.

Cells are stained with fluorescent dyes specific for a particular cell constituent (e.g. ethidium bromide at 0.1 mg/ml stains DNA, or a fluorescent antibody may be used). The stained cells are analysed as they pass in single file across sensing devices. Thus during transit (2–3 μs) across an argon ion laser beam a fluorescent light flash is emitted from each cell which can be analysed for both intensity and colour. In addition a cell volume signal may be generated either by placing a Coulter orifice in the cell stream or more simply, by measuring scattered light using a second photomultiplier tube set at right angles to the main beam. The dual signals from fluorescence and scattered light can be processed to give a three-dimensional picture of cell number, volume and DNA content.

Cell sorting is achieved by first breaking the stream of cells into droplets such that 1% of the droplets contain a single cell. Droplets are formed and pass between electrically charged plates a fixed time after analysis. Cells of particular characteristics (e.g. containing a G2 amount of DNA) can be sorted by giving the cell stream a charge for a short time such that three droplets (the centre one of which is calculated to carry the cell of interest) are deflected from the main stream into a separate collecting tube as they pass between the charged plates.

To give some idea of the results CHO cells pulse labelled with [³H]thymidine were stained with either ethidium bromide or using the acriflavin–Feulgen method (again staining DNA) and submitted to flow microfluorometry and cell sorting. Figure 10.13 shows the distribution of DNA content per cell. Approximately 61% of the cells had the G1 content of DNA and about 16.3% had the G2 + M content. The remaining 23% of the cells fell in between these two

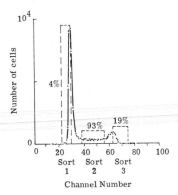

Fig. 10.13. Distribution of cells separated by flow microfluorometry. CHO cells were pulse labelled for 15 min with [³H]thymidine (1 μCi/ml) and stained with ethidium bromide. They were then submitted to flow microfluorometry and cell sorting on the basic of cellular DNA content. Cells from the indicated portions (sort 1, 2 and 3) were then subjected to autoradiography and shown to contain respectively 4%, 93% and 19% of the cells labelled. (Reproduced from Kraemer et al., 1973, with kind permission of the authors and publisher.)

values. Use is made of the formula derived by Lennartz and Maurer (1964):

$$dN/N_{(t)} = \left[\exp\left(\frac{1-\theta}{T}\right)\ln 2\right]d\theta \ln\frac{2}{T}$$

= the fraction of cells in the population which occupy the interval $d\theta$ of the cell cycle at any one time t.

It is this formula which gives Figure 10.5 and from which values for tG1 (8.4 h), t(G2 + M) (3.5 h) and tS (4.1 h) may be obtained.

The leading edge of the G1 peak of cells was shown by autoradiography to have only 4% of the cells labelled and similarly the trailing edge of the G2 + M population had 19% of the cells labelled. Cells sorted from between the two peaks had 93% labelled.

The sensitivity of the method is shown by the detection of a minor peak representing less than 1% of the cell population with a DNA content twice that of G2 + M cells.

A computer program has been developed (Dean and Jett, 1974) which resolves the flow microfluorometric distribution into G1, S and G2 + M subpopulations and enables cell cycle analyses to be performed rapidly and accurately. The cell sorting, however, takes place at a rate of 50,000 cells/min, i.e. it takes 20 min to sort 10^6 cells. Thus only biochemical analyses of high sensitivity can make use of this very expensive apparatus.

10.8. Drugs, radiation and the cell cycle

10.8.1. S-phase

RNA synthesis is required for entry of cells into S-phase but not for DNA synthesis itself, although S-phase is not completed in the presence of actinomycin D (a specific inhibitor of RNA synthesis). Addition of actinomycin D 2 h after the start of S-phase allows virtual completion of S-phase but cells do not pass through G2 and divide (Mueller and Kajiwara, 1966). It has been suggested that in order to replicate the last 3% of DNA RNA must be made late in S-phase.

Inhibition of protein synthesis has an immediate effect on DNA synthesis (Stimac et al., 1977). The rate of fork movement (i.e. the rate of progression of the replication fork) is reduced within 15 min and the extent of inhibition parallels the extent of inhibition of tritiated thymidine incorporation for up to 60 min. At longer times increased inhibition of DNA synthesis is a result of a decline in the frequency of initiation of new replicons. DNA synthesis is not as sensitive as is protein synthesis to inhibitors of protein synthesis, and this results in chromatin made in the presence of protein synthesis inhibitors being deficient in protein (White and Eason, 1973) and consequently more susceptible to nucleases (Weintraub, 1976).

Obviously inhibitors of DNA synthesis block progression of the cells through S-phase. Inhibitors include folic acid analogues (aminopterin and amethopterin; § 11.8.1 and § 13.2.1), fluoro-deoxyuridine (see § 11.8.2) and inhibitors of ribonucleotide reductase (hydroxyurea and high thymidine concentrations; see § 11.8.3 and § 11.8.4).

10.8.2. G2 phase

Log total cell number is a linear function of time for random log phase cultures. Therefore, if at zero time an inhibitor is added which blocks progress of the cells at a time t before division only those cells nearer than t to division will progress. The log of cell number will continue to increase linearly for a time t when the curve will flatten, thus defining the point in time of action of the inhibitor (Tobey et al., 1971).

Experiments using actinomycin D have shown a requirement for RNA synthesis about 30 min before M (Kishimoto and Lieberman, 1964; Arrighi and Hsu, 1965) and protein synthesis is required throughout G2 for cell division to occur as shown by the use of puromycin (Kishimoto and Lieberman, 1964). More accurate experiments in CHO cells put the requirement for RNA synthesis 1.8 h

before mitosis and showed the requirement for protein synthesis
ceased 0.6 h before mitosis (Tobey et al., 1966).

In a similar way the point in the cell cycle susceptible to low
doses of X-irradiation was found. 9 rads is a dose which does not kill
the cells but temporarily inhibits their rate of passage through G2
(Fig. 10.14). The point of action is 0.7–1.4 h before mitosis. X-irradiation of cells in this region stops their progress completely. However,
after 1.35 h the cells recover and join the unaffected cells in their
progress towards division so that from 2.05 to 3 h cells in irradiated
cultures accumulate in mitosis at a faster rate than do cells in
control cultures.

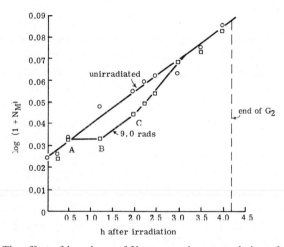

Fig. 10.14. The effect of low doses of X-rays on the accumulation of mitotic cells.
A similar experiment to that described in Fig. 10.9 for HeLa S3 cells, but 3 h after
colcemid addition half the cultures were irradiated with 9 rads X-rays (zero time on
the graph). At 0.6 h before prophase (A) there is a complete block in the progression
of cells towards mitosis. However, after a further 0.7 h (B) cells accumulate in mitosis
at a normal rate indicating that only cells 0.6–1.3 h before prophase are affected.
By 1.95 h (C) these cells recover from their inhibition and join unaffected cells in
their progress to division. ○, Control unirradiated cells; □, irradiated cells. (From
Puck and Steffen, 1963.)

10.8.3. G1 phase

RNA and protein synthesis are required for progression of cells through the G1 phase. Also, as discussed above (§ 10.6; see also § 2.3 and Chapter 15) a variety of situations including suboptimal growth conditions and stimuli to differentiation (e.g. the addition of dimethyl sulphoxide to cells of the Friend erythroleukaemia) lead to the cells stopping in G1 phase.

The use of mutants has indicated that RNA polymerase II activity is required midway through G1 in order that cells may enter S-phase (Rossini and Baserga, 1978).

10.9. Thymidine kinase and computer simulation of the cell cycle

Studies of the cell cycle have suffered from the disadvantage that there are too many variables to be able to accurately transform ideas of what may be happening into quantitative predictions. Thus there are many reports of the changes in the activities of several enzymes during the cell cycle and one might suggest that these changes are consistent with control at any one of a series of different levels. Perhaps thymidine kinase is the most studied enzyme (Adams, 1969a; Stubblefield and Mueller, 1965; Littlefield 1966a, b; Bello, 1974), but periodic changes have been reported for deoxycytidine kinase (Brent, 1971), thymidylate kinase (Brent et al., 1965), deoxycytidylate deaminase (Gelbard et al., 1969), ribonucleotide reductase (Turner et al., 1968; Murphree et al., 1969) and DNA polymerase α (Lindsay et al., 1970; Furlong et al., 1973).

In each case the activity rises throughout S-phase and falls sometime in late G2 or very early G1. Bello (1974) showed that the stability of thymidine kinase varied little throughout the cell cycle and that the onset of enzyme synthesis occurred at the time of onset but independently of DNA synthesis. Termination of synthesis required completion of S-phase and appeared to be dependent on RNA synthesis during G2.

A computer program 'Cell Sim' has been devised by C.E. Donaghey of the University of Houston and used by Stubblefield and Dennis (1976) to test the suggestion that the mRNA for thymidine kinase is made at a constant rate only during S and G2 phase and that it is immediately translated into active enzyme. The in vivo findings are consistent with this simulation for Chinese hamster fibroblasts of known cell cycle parameters if each cell makes 1 mRNA molecule per min during S and G2 and every 3 min each mRNA makes an enzyme molecule. The half-lives of the mRNA and enzyme molecules are 9 and 150 min, respectively. The simulation, however, fails to predict the rapid fall in activity in early G1 and the postulate is that the enzyme becomes less stable at this point. Whether this fits with Bello's finding (1974) of a small decrease in half-life from about 6.2 h to 4.4 h between early S and early G1 phases in KB cells was not ascertained by the initial computer program but should be feasible with Cell Sim II.

CHAPTER 11

Cell synchronisation

11.1. Introduction

Two different principles are used when a population of cells at a unique position in the cell cycle is required.

a. It is possible to block cells so that they accumulate at a specific stage in the cycle. This may be done chemically or physiologically but such methods suffer from the disadvantage that the cells have been interfered with and are therefore already abnormal in certain respects (see, however, § 11.7).

b. Cells at a particular stage of the cell cycle may be selected. This may be on the basis of some physical property, e.g. the loose attachment of mitotic cells or the varying size or DNA content of cells at different stages of the cycle. Alternatively, it may involve a chemical characteristic allowing remaining cells to be selectively killed.

Which method is chosen depends on the requirements of the experiment. It is sometimes preferable to perform the initial stages of an experiment (e.g. isotopic labelling) with an unsynchronised exponentially growing culture and then to select cells at a given phase of the cycle by e.g. zone centrifugation for enzymic analysis. In such experiments the cells, following selection, are not generally replaced in culture and so they may be exposed to an otherwise deleterious environment.

11.2.   Selection of mitotic cells

Cells growing as a monolayer round up when they divide. They therefore are less firmly attached to the substratum at this time and may be easily detached (Terasima and Tolmach, 1961, 1963; Robbins and Marcus, 1964; Peterson et al., 1968; Shall, 1973a). Exponentially growing cells are used and it is important to maintain constant pH and temperature throughout the selection procedure.

Klevecz et al. (1974, 1975) have used CHO cells growing in McCoy's medium with 20% foetal bovine serum and Hepes buffer. They maintain the cells throughout in this medium at 37°C and of the cells they select 98–99% are in mitosis. The viability of these selected cells approaches 100% and on reseeding half the cells attach within 1 h of selection and maximum attachment is found by 4 h (Klevecz, 1975). However, they select a very small proportion of the original cells. As different cells attach with differing firmness to the substratum, the following procedures may need to be modified for each cell type.

11.2.1. Shaking

– Inoculate cells into bottles 24 h prior to initial selection. 5×10^7 Don-C Chinese hamster fibroblast cells may be inoculated into a roller bottle in McCoy's medium (see § 11.2.2).
– After 24 h shake the bottle of cells (at least 10^7 cells should be present initially) and discard the medium which contains dead cells and cell debris. The shaking is best done in a shaking water bath so that the bottle is shaken 20 times in 3 s with the medium washing over the cell monolayer. The firmness of attachment of different cell lines differs considerably and the vigour required to dislodge mitotic cells must therefore be investigated anew for each case.
– Add fresh prewarmed medium (take care not to pipette it onto the cell monolayer) and incubate for further periods of 15, 30, 45 min, etc. The actual intervals between harvests vary from cell line to cell line.

- Harvest loosely attached cells by shaking as above and this time collect the medium on ice, pooling all batches.
- Sediment the cells at 700 g for 3 min.
- Reseed the mitotic cells at about 10^5 per 5 cm dish, when they should attach and divide within an hour.

Although the yield of mitotic cells is poor (1–3%) it is increased by pooling many harvests taken over a 2–3-h period.

Another way of increasing the yield of mitotic cells is to accumulate cells in metaphase by a brief treatment with colcemid. By subculturing Don-C Chinese hamster fibroblasts every 24 h Stubblefield et al. (1967) achieved a partial synchronisation with mitotic peaks at 7 and 19 h after subculture. By adding colcemid at 18 h and selecting mitotic cells at 22 h after subculture they obtained a yield of 21% of the population, 86% of which were in metaphase. These cells grew normally and exhibited no significant deviations from control cultures in their mitotic interval, generation time, DNA synthesis kinetics or proliferative capacity. This method, however, is not universally applicable as not all cell lines show reversible colcemid inhibition.

11.2.2. Trypsinisation

Stubblefield et al. (1967) used dilute trypsin to release their accumulated metaphase cells and claim that the purity of the resulting suspension was better than with shaking. The details of their method are:

- Subculture Don-C cells every 24 h in McCoy's 5a medium (Appendix 1) containing 20% foetal calf serum and 0.08% lactalbumin hydrolysate, seeding cells at 1.2×10^5/ml (3×10^4 cells/cm²).
- 18 h after subculture add colcemid (0.06 μg/ml) and continue incubation for 4 h. Some mitotic cells float free at this stage and should be combined with those released later by trypsinisation.
- Remove the medium and replace with 0.1% cold trypsin in PBS.
- Shake the cultures gently (about 2 cycles/sec) either by hand or in a shaking water bath.

- After 45 s transfer the suspended mitotic cells to centrifuge tubes, pellet at 400 g for 2 min, and resuspend in fresh medium at 4×10^4 cells/ml.
- After 20–30 min the cells divide synchronously. Again this method is not universally applicable as not all cell lines show reversible colcemid inhibition.

11.3. Selective killing of cells in particular phases

This is usually done either with high specific activity tritiated thymidine (Whitmore and Gulyas, 1966) a method which is of fairly general applicability or with hydroxyurea (Sinclair, 1965), a method applicable only to certain susceptible cells.

The aim is to kill cells in a particular phase of the cell cycle, thereby selecting for growth those cells outwith that phase. As the cells are growing exponentially and passing round the cell cycle, the longer the period of exposure the fewer survivors and the narrower the width of the surviving population. This may be true in theory, but in practice if the killing is allowed to remove 90% of the cells, those remaining rather than being concentrated just before S-phase have in fact selected themselves as slow growers, and it may take several hours before they enter S-phase, and by the time they enter G2 most synchrony has been lost.

Moreover a major disadvantage with such methods is that the dead cells, although unable to divide, may continue to metabolise for a considerable period, thus confusing any biochemical investigation.

Cells in mitosis may be selectively removed and discarded over a long period of time. This leaves behind a population most of which should be in G2-phase (Stubblefield, 1964; Creasey and Markiw, 1965; Pfeiffer and Tolmach, 1967). Unfortunately, as well as the probability of selecting slow growers, the use of colchicine, colcemid and vinblastine may not only inhibit mitosis but may also have other effects on cells such as inhibition of RNA synthesis.

11.4.   Selection of cells by size

11.4.1.   Electronic cell sorting

This is considered in more detail in § 10.7.5. It is a method which involves a modified electronic cell counter, i.e. a Coulter counter linked to a pulse height analyser and an electronic cell sorter. It is capable of separating cells at a rate of about 50,000/min into a number of size classes. The cells remain viable and show no change in generation time (Fulwyler, 1965; Van Dilla et al., 1967). However, because of the high cost of the instrumentation and the relatively slow sorting rate this method has found limited applications in biochemistry at present. As an alternative to the use of the electronic cell counter, cell size may be analysed by measuring scattering of light from a laser (§ 10.7.5).

11.4.2.   Zone sedimentation

The rate at which cells sediment depends on their density and size. It is therefore relatively easy to separate cells of different types which differ in density but more difficult to separate cells of the same type (but at different stages of the cell cycle), as in this case the rate of sedimentation (mm/h) is roughly equal to $r^2/4$ where r is the cell radius in μm. A two-fold increase in volume (the maximum to be expected) results in only a 1.6-fold increase in sedimentation rate.

In order to stabilise the sedimenting cells, it is necessary to include a gradient and those most commonly used are serum or Ficoll (Pharmacia Ltd) (Boone et al., 1968; Miller and Phillips, 1969; Macdonald and Miller, 1970; Warmsley and Pasternak, 1970). Sucrose has also been used (Sinclair and Bishop, 1965; Morris et al., 1967; Shall and McClelland, 1971; Shall, 1973b) but tends to lower cell viability, unless great care is taken to maintain isotonicity.

Boone et al. (1968) centrifuge cells through a discontinuous 10–20% Ficoll gradient made up in Eagle's minimum essential medium modified for suspension (i.e. lacking calcium and bicarbonate and containing 10 times the normal phosphate concentration). They use an A-1X zonal centrifuge rotor and spin for 1 h at 1000 r.p.m. at 20°C, and obtained clear separation of different cell types (HeLa and rabbit thymocytes).

Sinclair and Bishop (1965) and Morris et al. (1967) also use a centrifuge to increase the rate of sedimentation, but Shall (1973b) and Miller and Phillips (1969) use sedimentation under unit gravity and both describe simple pieces of apparatus for construction of the gradients and obtaining the separation. Although Shall claims up to 10^9 cells may be separated at a time, the maximum that can be separated on the Miller and Phillips apparatus is 3×10^8. The two pieces of apparatus are essentially the same and are shown dia-grammatically in Figure 11.1. If made of glass the apparatus may be sterilised by autoclaving (cover the open ends of the tubes and reservoirs with aluminium foil) and the cells used for further growth.

The gradient is formed from two solutions. A_1 contains either (a) 10% w/v sucrose in complete medium which has the NaCl concentra-tion reduced by 146 mM to maintain constant osmotic pressure, or (b) 30% foetal calf serum in PBS; A_2 contains either (a) 2.72% w/v sucrose in complete medium which has the NaCl concentration reduced by 40 mM or (b) 15% foetal calf serum in PBS. The volumes,

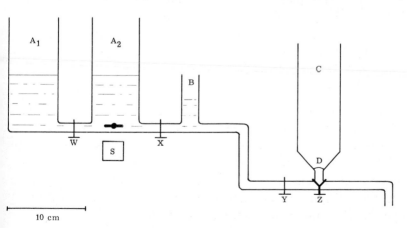

10 cm

Fig. 11.1. Apparatus for separating cells by sedimentation under unit gravity (see text for description).

and hence heights, of the two solutions must be the same (about 250 ml).

- With tap X closed, open W and turn the bar magnet using a magnetic stirrer. The magnet should rotate rapidly to ensure thorough mixing of the components.
- Into B place 10–30 ml of (a) serum free medium or (b) PBS and allow it to run into C to remove air bubbles.
- Into B add the cell suspension (10–20 ml) in growth medium containing 10% foetal calf serum and allow this to flow into C where it will form a narrow band beneath the medium or PBS already there.
- Now, without letting air into the system, allow the gradient to flow from A into C. The baffle (D) prevents the solutions squirting into C, thereby disturbing the cell layer. About 500 ml of gradient should be added at 30–40 ml/min. The rate of flow can be controlled by a 'Rotaflow' tap Y (Quickfit and Quartz Ltd). C should contain about 100 ml for every cm of height.
- Allow to stand for 2–4 h. The lower the temperature, the greater the viscosity and hence the slower the sedimentation. However, if the sedimentation time is long compared to the cell generation time, it is better to stand at 4°C. If too many cells are placed on the gradient 'streaming' may occur.
- Using the three-way tap – Z – allow the separated cells to flow out of C. Discard the medium from the conical part of C and then collect 15 ml fractions.

Macdonald and Miller (1970) describe an even simpler apparatus and Shall (1973b) describes a mini version where the gradient (15 ml) is formed in a 15 × 1.6 cm test tube, and the cell suspension (3 × 10^7 cells) is layered onto this with a wide bore pipette. After standing upright for 50 min at 37°C the topmost 1 ml which contains the smallest and youngest cells may be removed, the sucrose washed off and the cells returned to normal growth medium.

This method allows the separation of large numbers of cells in a short time and the cells show reasonable viability. However, the degree of synchrony is only moderate compared with selection of mitotic cells.

11.5. Synchronisation by subculture

It is often difficult to avoid introducing some degree of synchronous growth in a culture simply as a result of routine operations. Thus Stubblefield et al. (1967) found that subculture of Don-C cells every 24 h exerted a selection pressure favouring cells with a 12 h generation time and highest mitotic frequencies at about 7 and 19 h.

More frequently a cell culture is allowed to leave the exponential growth phase and enter a stationary phase before it is subcultured (see § 5.2). When this happens the subsequent round of DNA synthesis and cell division is partially synchronised.

– Establish cultures of mouse L929 cells and feed every two days with Eagle's medium (Glasgow modification) supplemented with 10% calf serum (Fig. 5.2).
– After 10 days subculture the cells initiating new cultures at 2×10^5 cells/ml. These cultures may be on coverslips for subsequent autoradiographic analysis or in Roux bottles for biochemical experimentation.
– Label coverslip cultures with tritiated thymidine (5 μCi/ml; 5 Ci/mmol) for 30 min periods at various times after subculture. This is conveniently done by adding to 0.5 ml medium 10 μl of a solution containing 2.5 μCi [^{3}H]thymidine at 50 μM.
– Process the cells for autoradiography (§ 12.3.1) and score the percent of cells labelled and the mitotic index.

It is found (Fig. 11.2) that few cells enter S-phase before 12 h, but by 18 h about 70% of the cells will be making DNA. All the cells divide shortly after 24 h. This is somewhat longer than the normal cycle time and this has been interpreted in two different ways, viz.:

1. Subculturing provides a stimulus and, after a short lag the cells begin to traverse the whole of G1 before passing through S-phase to mitosis.

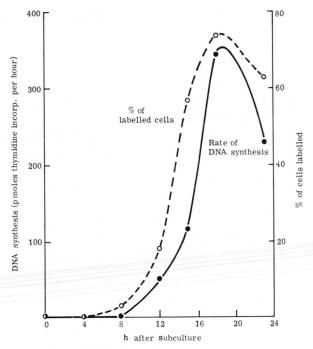

Fig. 11.2. DNA synthesis in cells subcultured from stationary phase culture. Mouse L929 cells were subcultured into 5 cm dishes (4×10^5 cells/5 ml Eagle's MEM containing 10% calf serum). At the indicated times they were incubated for 1-h periods with tritiated thymidine and incorporation into DNA measured.

2. There is a period of about 12 h while cells are growing, and only then is their transition probability increased sufficiently to allow them to pass from the A state to the B phase and initiate DNA synthesis (see § 10.4).

Although limitations of space are important in causing the cells in crowded cultures to cease growing limitations on the supply of nutrients are also important (Dulbecco and Elkington, 1973) and form the bases of the next two methods.

11.6. Serum deprivation

Burk (1970) showed that Syrian hamster cells (BHK21/C13) failed to grow when transferred to medium containing 0.25% serum and could be maintained in a quiescent state for 8 days or more. On readdition of serum no DNA synthesis occurred for 9 h and mitotic peaks were observed at about 23 and 33 h. It appeared the cells had come to rest in G1 and on stimulation showed a lag of about 9 h before entering into exponential growth with a generation time of about 10–12 h. Adding serum for just 3 h (serum pulse) induces about 50% of the cells to pass through one cycle.

Temin (1970) and Todaro et al. (1965) showed similar effects for chicken fibroblasts and 3T3 mouse fibroblasts.

The proportion of committed cells depends almost linearly on the duration of the serum pulse (Brooks, 1976). The presence of nucleosides potentiates the serum response with the result that the rate of entry into S-phase increases abruptly after the lag period in a manner, suggesting commitment is a random event occurring with a fixed probability (Brooks, 1975) similar to that proposed by Smith and Martin (1973, 1974).

The recommended procedure for synchronisation of hamster cells is:

– Suspend BHK21/C13 cells at $10^6/10$ ml Eagle's Glasgow modification supplemented with 10% tryptose phosphate, 0.05 mM L-serine, 0.1 mM L-ornithine, 0.1 mM hypoxanthine and 0.25% calf serum and plate into 5 cm dishes. With this parent cell line very little DNA synthesis or cell growth occurs, but with the polyoma transformed PYY cells two cell doublings occur in low serum.
– After 100 h add 0.5 ml calf serum per plate to initiate cell growth.

11.7. Isoleucine starvation

It was observed that CHO cells which had entered a stationary phase in Ham's F10 medium could be stimulated to undergo a further round of division by changing the medium. The growth limitation was not a serum factor but was traced to a deficiency in isoleucine which is present in Ham's F10 medium at only about 5–10% its concentration in other media (Ley and Tobey, 1970; Tobey and Ley, 1970).

Exponentially growing cells (CHO, L929, BHK21/C13) can be transferred to isoleucine deficient medium containing dialysed serum when they become arrested in G1 phase (Tobey and Ley, 1971; Everhart, 1972; Tobey, 1973; Skoog et al., 1973). The rate of DNA synthesis falls for several hours after transfer to isoleucine deficient medium and the cell number rises by about 30% as those cells in S, G2 and M divide. (Small amounts of isoleucine allow a somewhat larger increase in cell number; Tobey, 1973.) Imbalanced growth does not appear to occur and biosynthetic capacities are maintained at much higher levels than in stationary phase cultures (Enger and Tobey, 1972).

Readdition of isoleucine leads to induction of DNA synthesis in most cells starting at 4–5 h and the mitotic index begins to increase at 12 h in CHO cells.

- Harvest an exponentially growing suspension of CHO cells and wash them in isoleucine free Eagle's medium.
- Plate 7.5 × 10⁵ CHO cells into 5 cm dishes in Eagle's MEM lacking isoleucine and containing 5% dialysed foetal calf serum.
- After overnight incubation replace the medium with fresh medium containing isoleucine (2×10^{-6} M) and 10% undialysed foetal calf serum.
- If cells are plated onto coverslips [³H]thymidine pulses (5 μCi/ml) will show the appearance of S-phase cells 4 h after isoleucine addition.

The mechanism of action of isoleucine starvation is not clear but it is not simply a general deficiency of amino acids as e.g. leucine starvation causes a far more drastic inhibition of cell growth without any synchronising action (Everhart, 1972; Tobey, 1973). Rather it may indicate some subdivision of G1 phase.

This method gives results very comparable to selection of mitotic cells in that almost 100% of the population can be obtained in G1. It is much simpler than mitotic selection and is more readily applied to larger numbers of cells.

11.8. Blockade of S-phase

Synchrony at the G1/S interphase has been accomplished by interfering with the synthesis of one or more deoxyribonucleoside triphosphates, which are required for DNA synthesis while allowing other cellular procedures such as synthesis of RNA and protein to proceed.

If the block is maintained for about 16 h (i.e. $T-tS$) about 70% of the cells of an exponentially growing culture gather at the beginning of S-phase. In practice it is difficult to know whether the cells actually do make some DNA but at a greatly reduced rate, i.e. whether the cells have entered S-phase or not. Those cells which were caught in S-phase when the blockade was imposed will remain there until the block is reversed when the 70% of the cells at the G1/S border pass as a synchronised cohort through S-phase and G2 to division. These synchronised cells are often very suitable for studying the events occurring in the S and G2 phases of the cell cycle. However they suffer from two disadvantages.

1. Only 70% of the cells are in synchrony – the remaining 30% being up to 6 h ahead of the main group.

2. Events other than those concerned directly with DNA synthesis may proceed in the presence of reagents which block DNA synthesis and certain aspects of the cell's metabolism may reflect G2 or even G1 activity while DNA synthesis related events are held at the G1/S phase boundary.

This unbalanced growth rapidly leads to cell death if prolonged for more than a generation time (Ruekert and Mueller, 1960). Unbalanced growth will occur in any cells committed to division (§ 10.4) yet blocked in one function including those maintained in colcemid for more than a few hours. Moreover, selective blocking of DNA synthesis may have effects which are not apparent until the synchronised cells are released and proceed to the next G1 phase (Firket and Mahieu, 1966; Cress and Gerner, 1977). Schindler et al. (1968) have shown that the duration of G2 is independent of the length of time cells are blocked with amethopterin up to 8 h.

The first disadvantage, however, can be overcome by using a double block technique. In this treatment the agent which blocks DNA synthesis is added to cells which have already been treated or selected so that all the cells are in the G1 phase. This ensures collection of all the cells at the G1/S boundary. The initial selection may be brought about by selecting mitotic cells (see § 11.2) or by using a population of cells which have just been induced into a growth phase by addition of serum (§ 11.6) or a limiting amino acid (§ 11.7), by subculture (§ 11.5) or by addition of phytohaemagglutinin (§ 6.2.5) for example. The term 'double block technique', however, is usually applied to cells where both selection and imposition of the G1/S block are by a similar method. Thus high concentrations of thymidine may be used to block DNA synthesis (see § 11.8.3). If a population of cells is treated with 3 mM thymidine for 16 h and then the thymidine removed the 70% of cells arrested at the G1/S phase boundary, together with the 30% of cells arrested in S-phase will move around the cell cycle such that 8–10 h later all the cells will be in a sector of G1. If the thymidine block is now reimposed within about 12 h all the cells will have moved to the beginning of S-phase where they will have been arrested.

In practice synchronisation at the G1/S boundary may be achieved by a number of reagents as well as thymidine, e.g. aminopterin, amethopterin, 5-fluorodeoxyuridine or hydroxyurea.

11.8.1. Action of aminopterin and amethopterin (methotrexate)

On addition to cells these inhibitors rapidly bring about inhibition of DNA synthesis. Incorporation of [³H]deoxyuridine into DNA is reduced to 3% of control values within 15 min of addition of 10 μM aminopterin and even 0.1 μM reduces incorporation to less than 20% of controls though this takes 45 min (Siegers et al., 1975). Although concentrations of amethopterin between 0.1 and 1 μM block synthesis of thymidylate, DNA synthesis continues at a significant rate and pools of thymidylate are slow to dissipate (Adams et al., 1971; Baumunk and Friedman, 1971; Fridland, 1974; Fridland and Brent, 1975). Aminopterin and amethopterin are 4-amino analogues of folic acid (Fig. 13.3) and as such are potent inhibitors of the enzyme dihydrofolate reductase (EC 1.5.1.3) (Blakley, 1969). This enzyme catalyses the reduction of folic acid and dihydrofolic acid to tetrahydrofolic acid which is the level of reduction of the active coenzyme involved in many different aspects of single carbon transfer. Thus, as described in Chapter 13, tetrahydrofolate is involved in the metabolism of a) the amino acids glycine and methionine; b) the carbon atoms at positions 2 and 8 of the purine ring; c) the methyl group of thymidine; and d) indirectly in the synthesis of choline.

In order to concentrate on the effect of the drugs on thymidine metabolism the other actions are bypassed by addition to the culture medium of hypoxanthine or adenosine (30 μM or 200 μM) and glycine (100 μM) in addition to the methionine and choline normally present. Cells possess considerable pools of tetrahydrofolate, and, except during synthesis of the coenzyme from the vitamin, the only time the enzyme dihydrofolate reductase is required is to regenerate tetrahydrofolate from the dihydrofolate during synthesis of dTMP, i.e. during S-phase. Thus only in S-phase cells will the drug have the effect of depleting the pool of tetrahydrofolate and cause the accumulation of dihydrofolate. However, no accumulation of dihydrofolate was found when Chinese hamster cells were grown for 24 h in medium containing 0.25 μM amethopterin (McBurney and Whitmore, 1975) and concentrations of amethopterin up to 2.5 μM did not affect glycine synthesis except in cells previously starved of folic acid for 48 h (McBurney and Whitmore, 1975), i.e. in cells lacking a pool of tetrahydrofolate. This led McBurney and Whitmore to suggest a more direct effect of amethopterin on thymidylate and purine synthesis and an action of the drug on thymidylate synthetase has been reported (Borsa and Whitmore, 1969). McBurney and Whitmore (1975) do show that reduction of folic acid is more sensitive to amethopterin than is inhibition of thymidylate synthesis, and on exposure to low doses (0.02 μM) cell growth would become limited by lack of active folate reductase after 48 h. It was by prolonged exposure to low doses of amethopterin that the mutant lines were selected and this could reconcile the two sets of data.

The inhibitory action of the antifolates may be reversed either by changing the medium (Adams, 1969a, b), or more conveniently by addition of thymidine to the medium when dTTP is synthesised using the enzymes thymidine and thymidylate kinase.

11.8.2. Action of 5-fluorodeoxyuridine

5dFUrd is an analogue of thymidine and is taken up and phosphorylated in a similar manner to form 5dFUMP which is a competitive inhibitor with dUMP of thymidylate synthetase. However, as it may compete with thymidine for uptake and phosphorylation it is not recommended for quantitative studies of DNA synthesis. Its action may be reversed in a similar manner to that of the antifolates.

11.8.3. Action of high concentrations of thymidine

Thymidine is taken up by cells and rapidly converted to dTTP, the pool size of which is related to the extracellular thymidine concentration (see Fig. 12.5). At thymidine concentrations as low as 3×10^{-7} M this leads to a measurable effect on the rate of DNA synthesis (Cooper et al., 1966). At concentrations above 1 mM inhibition of DNA synthesis is almost complete (Morris and Fischer, 1960; Xeros, 1962; Bootsma et al., 1964; Studzinski and Lambert, 1969). Bostock et al. (1971), however, claim that DNA synthesis proceeds at about one-third the normal rate even in the presence of 2 mM thymidine, and it is obvious from Figure 12.3 that 5 mM thymidine is required to stop cell growth.

Inhibition of DNA synthesis is brought about by the action of dTTP as an allosteric inhibitor of ribonucleotide reductase (Reichard et al., 1961; Moore and Hurlbert, 1966; Brown and Reichard, 1969; Kummer et al., 1978). This enzyme is responsible for reducing all four ribonucleoside diphosphates (NDP) to the corresponding deoxyribonucleoside diphosphates (dNDP). It is subject to a complex allosteric control which has been most studied with the bacterial enzyme. Most studies with the mammalian enzyme show it to be similar to the bacterial enzyme, but recent work by Peterson and Moore (1976) and Cory et al. (1976) may point to the presence of two enzymes each catalysing the reduction of different substrates.

dTTP is required as an allosteric activator for the reduction of GDP, and, in turn, dGTP is an allosteric activator for reduction of ATP. However, dTTP is an allosteric inhibitor for reduction of CDP and UDP whereas dATP is an allosteric inhibitor for reduction of all four NDPs. It is this complex system which is believed to control the balanced supply of dNTPs for DNA synthesis. As reduction of UDP or CDP is on the pathway to dTTP production control at this point is not unexpected (Fig. 11.3) and DNA synthesis is inhibited by high concentrations of dTTP reducing the supply of dCTP. Apart from removing the inhibitory concentrations of thymidine by medium change the block may also be reversed by addition of deoxycytidine at 10 μM (Morris and Fischer, 1960; Bjursell and Reichard, 1973).

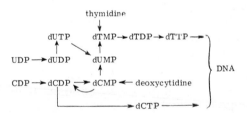

Fig. 11.3. Interconversions of deoxyribonucleotides.

11.8.4. Action of hydroxyurea

Hydroxyurea also inhibits DNA synthesis by its action on ribonucleotide reductase, but in this case it is the reduction of the purine nucleoside diphosphates which is inhibited and the pool of dTTP rises slightly (Turner et al., 1966; Adams and Lindsay, 1967; Krakoff et al., 1968; Adams et al., 1971; Skoog and Bjursell, 1974). Its action is most satisfactorily reversed by changing the medium for drug free medium.

11.9.   Procedure for inducing synchrony at the G1/S interphase

11.9.1.   Isoleucine starvation and hydroxyurea

– Plate 7.5 × 10^5 washed CHO cells into 5 cm dishes in Eagle's MEM medium lacking isoleucine and containing 5% dialysed foetal calf serum.
– After 8 h incubation change the medium for Eagle's MEM containing isoleucine, 10% undialysed foetal calf serum and 2 mM hydroxyurea.
– After 16 h change the medium again for complete medium lacking hydroxyurea. Most of the cells will be at the G1/S interphase and will start to replicate DNA immediately. This may be followed autoradiographically; by total incorporation of [^{3}H]thymidine; and later by the appearance of mitotic cells.

11.9.2.   Stationary phase cells and aminopterin

– Establish a Roux bottle of mouse L929 cells in Eagle's Glasgow modification containing 10% calf serum.
– Change the medium on every second day.
– On day 8 trypsinise the cells and plate 4 × 10^5 cells into 5 cm dishes in the above medium supplemented with aminopterin (2 μM) adenosine (200 μM) and glycine (100 μM).
– After 16 h reverse the inhibition by addition of thymidine (2 × 10^{-5} M) and deoxycytidine (2 × 10^{-6} M) when the cells will begin to start making DNA, slowly at first but at maximum rate 3 h later.

11.9.3.   Double thymidine block

– Plate 3 × 10^5 HeLa cells in 5 cm dishes in Eagle's Glasgow modification containing 10% calf serum and 3 mM thymidine.
– After 16 h change the medium for fresh medium lacking thymidine.
– After a further 8 h add thymidine to a final concentration of 3 mM.
– After a further 16 h repeat the second stage when the cells will be released into S-phase synchronously.

11.9.4. Comparison of the methods

The double thymidine block is the method mostly referred to in the literature but it is tedious to apply and suffers from the disadvantage that the cells enter S-phase with a pool of dTTP which is decreasing over the first hour or so and this makes estimation of the rate of DNA synthesis difficult (Adams, 1969b).

The isoleucine/hydroxyurea method also involves two medium changes but the first of these may be replaced by simply adding isoleucine to the deficient medium.

The stationary phase and aminopterin method suffers from the disadvantage that it takes a week to prepare the cells but thereafter it is straightforward.

Variations and other combinations of these methods are obviously possible, e.g. mitotic selection and thymidine block; aminopterin block reversed with low thymidine followed by a high thymidine block reversed with deoxycytidine.

All these methods succeed in accumulating 80–90% of the cells at the G1/S interphase and are very suitable for obtaining populations of S-phase cells for study. They are readily scaled up to the level of roller bottles though at this scale the amount of thymidine required for a double block is considerable. Concentrated solutions of thymidine can be made up and sterilised by autoclaving but most of the other solutions should be filter sterilised.

11.10. Synchronisation in G2

The G2 phase of the cell cycle is perhaps the most difficult to study as it is the most difficult phase in which to obtain a synchronised cell population. This is because, if cells are synchronised by selection at mitosis or accumulation at the G1/S boundary, by the time they reach G2 much of the synchrony has been lost. This is because of the dispersion forces arising from the different rates at which individual cells in a population traverse the cycle. G2 populations are always contaminated with cells in other phases of the cycle and the maximum fractions of Chinese hamster (CHO) cells obtainable in G2 are 0.7 by double thymidine block and 0.4 by mitotic selection (Enger et al., 1968).

Although cultured cell populations come to rest in G1, i.e. between cell division and DNA synthesis, a proportion of mouse ear epidermal cells are thought to be arrested in G2 (Gelfant, 1959, 1963), although recent evidence casts doubt on these conclusions (Sauerborn et al., 1978). There is some evidence that human embryonic fibroblasts maintained in culture for 48 passages (i.e., in the terminal phase) may arrest in G2 (Maciero-Coelho et al., 1966) but these are so abnormal as to be of no value in studies of G2.

One way of obtaining a population of G2 cells is to have a mutant with a temperature sensitive stage in G2 phase, and Basilico (1978) has described a procedure which should select for such temperature sensitive mutants.

CHAPTER 12

Use of radioactive isotopes in cell culture

Cell cultures offer many advantages over intact animals when it comes to incorporation of radioactive tracers and studying the effects of drugs or hormones. Thus the tracer or drug may be added and removed at known times, its extracellular concentration and specific activity maintained constant and there is no interference in its metabolism by cells of other organs. This does not mean to say, however, that there are no pitfalls to the use of isotopes in cell cultures.

12.1. Estimation of rates of DNA synthesis

Estimates of the rate of DNA synthesis in a cell population or in individual cells may be required as a measure of the rate of cell growth, or for cell cycle studies, or to satisfy a basic interest in DNA metabolism. However, on addition of tritiated thymidine to cells in culture a number of problems arise.

1. For incorporation into DNA the thymidine must be phosphorylated stepwise:

$$\text{Thymidine} \rightarrow \text{dTMP} \rightarrow \text{dTDP} \rightarrow \text{dTTP} \rightarrow \text{DNA}$$

Thus the cell does not need to have a permeability barrier and the various kinases must not be rate limiting if the incorporation is to reflect the rate of DNA synthesis rather than the activity of an intermediate rate limiting enzyme. However, the thymidylate kinases are absent from non-growing cells and thymidine kinase varies dramatically throughout the cell cycle (Stubblefield and Dennis, 1976). Thus uptake and phosphorylation may well be rate limiting, except during S and G2 phase, and evidence suggests that there is little uptake outside these phases (Adams 1969a). During S-phase, however, the kinases appear not to be rate limiting and analysis of the acid soluble pool shows it to be predominantly dTTP (Gentry et al., 1965; Adams 1969a).

2. Variations in the concentration of thymidine added to the medium affect its uptake, its phosphorylation and the pool size of

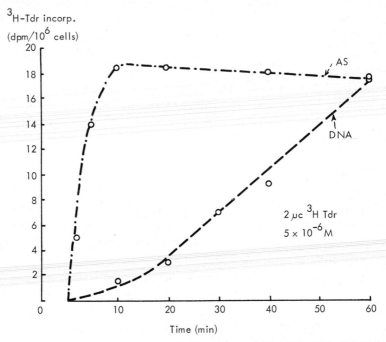

Fig. 12.1. Equilibration of tritiated thymidine with the acid-soluble pool and incorporation into DNA. Mouse L929 cells, growing in 5 cm dishes, were incubated with [^{3}H]thymidine (0.7 μCi/ml; 5 μM) for the indicated times after which the cells were quickly washed three times with ice-cold BSS. The acid-soluble material (AS) was extracted into cold 5% TCA and after further acid and ethanol washes the cells were solubilised in 0.3 N NaOH and incorporation into DNA measured. (Reproduced from Adams, 1969a, with kind permission of the publisher.)

dTTP. In fact within 10 min of addition of tritiated thymidine to a cell culture the amount of ^{3}H in the acid soluble pool (intracellular dTMP, dTDP and predominantly dTTP) reaches an equilibrium level (Fig. 12.1) (Gentry et al., 1965; Adams, 1969a). Although there is obviously a through put such that [^{3}H]dTTP is being formed and removed by DNA synthesis it appears that the size of the radioactive deoxythymidylate pool is determined by the concentration of extracellular thymidine (Fig. 12.2). This may come about by a combination of forward promotion of thymidine kinase by its substrate and feedback inhibition by dTTP (Ives et al., 1963). The

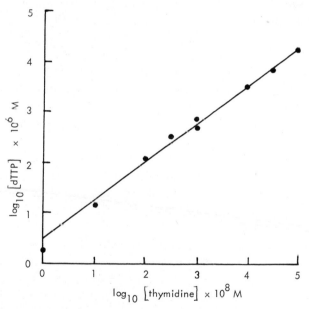

Fig. 12.2. Relationship between the concentration of extracellular thymidine and the exogenously derived nuclear pool of dTTP. Mouse L929 cells were incubated for 30 min with [^{3}H]thymidine at the concentrations indicated and the acid-soluble pool extracted (see Fig. 12.1). The concentration of dTTP is calculated on the basis that the radioactive acid soluble pool is completely [^{3}H]dTTP and that it is located in the nuclei of S-phase cells. The volume of a nucleus is taken as 500 μm^3. (Reproduced from Adams, 1969a, with kind permission of the publisher.)

size of the pool remains constant once equilibrium has been reached until extracellular thymidine levels begin to fall. This will happen within a few hours when the concentration of extracellular thymidine in 10^{-6} M and even at 10^{-5} M 10% may be utilised within 24 h. Thus, if 10^6 cells are doubling every 24 h they will require approximately:

$$10^6 \times \frac{10}{3} = 3.3 \times 10^6 \text{ pg thymidine}$$

$$= 10^4 \text{ pmol thymidine}$$

This means that if the only source of thymidine is that supplied in the medium 5 ml of 10^{-6} M thymidine which supplies 5×10^3 pmol is sufficient to support 10^6 cells growing exponentially for less than 12 h or

$$t = \frac{Tcv}{DN}$$

where t = time in hours until thymidine exhausted; D = DNA content in pg/cell; N = number of cells in millions; v = volume of medium in ml; c = thymidine concentration in nmol/ml (μM); T = generation time (h).

The pool appears to expand indefinitely as the extracellular concentration of thymidine increases up to 10 mM (Fig. 12.2), which shows that the kinases are present in vast excess (Cleaver and Holford, 1965; Gentry et al., 1965; Cooper et al., 1966; Adams, 1969a; Stimac et al., 1977).

3. dTTP is also synthesized endogenously. The pathway involves ribonucleotide reductase. This enzyme is subject to allosteric control and one of the controlling elements is dTTP which at high concentrations inhibits the reduction of CDP and UDP thus leading to a fall in the pool size of dCTP. This, in turn, leads to inhibition of DNA synthesis (Fig 12.3), an inhibition which may be reversed by addition to the growth medium of deoxycytidine at 5–10 μM (Fig. 12.4). This inhibition is detectable when thymidine is added to the grown medium at about 10^{-6} M and becomes absolute above 3 mM, and forms the basis of one method of synchronising cells (see Chapter 10).

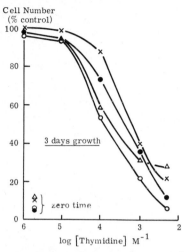

Fig. 12.3. Inhibition of cell growth by thymidine. About 10^5 cells in 5 ml Eagle's MEM, supplemented with extra vitamins and calf serum, were inoculated into 5 cm dishes and incubated in the presence of the indicated concentrations of thymidine. After 4 days the cells were trypsinised and counted using a Coulter counter. ●, L929; ○, HeLa, BHK; △, BSCl; ×, CHO.

4. The endogenously synthesised dTTP dilutes the specific activity of the [³H]dTTP formed from the added [³H]thymidine. Thus on adding tritiated thymidine at 3×10^{-8} M most of the DNA thymine is synthesised by the endogenous or de novo pathway, but when the [³H]thymidine concentration in the medium is raised to 0.3 mM it contributes 90% or more of the DNA thymine (Cleaver and Holford, 1965; Cooper et al., 1966; Cleaver, 1967). As the specific activity of [³H]dTTP is one of the factors which determine the amount of radioactivity incorporated into DNA (either total counts/min or grain counts) and as this varies (a) with external thymidine concentration and (b) with the state of the cells, the quantitative estimation of rates of DNA synthesis is full of pitfalls.

5. The concentration of endogenously synthesised dTTP varies during the cell cycle. Thus, in resting cells and cells in the G1 phase the pools of dTTP are low (about 3 pmol/10^6 cells) but these increase throughout S phase and G2 reaching a maximum at mitosis (Adams

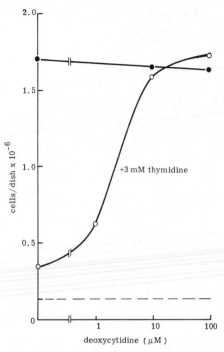

Fig. 12.4. Reversal of thymidine inhibition by deoxycytidine. CHO cells were inoculated into 5 cm Petri dishes (0.13 × 10⁶ cells per 5 ml Eagle's MEM containing twice the normal concentration of vitamins and supplemented with 10% calf serum and 57.5 mg proline/l). After 4 days the cells were harvested and counted using a Coulter counter. ●–●, Control; O–O, thymidine (3 mM) present throughout the incubation. The indicated concentration of deoxycytidine was present in both control and experimental dishes. The dashed line represents the initial cell density.

et al., 1971; Skoog et al., 1973; Walters et al., 1973). This dTTP pool is largely nuclear (Adams, 1969a; Adams et al., 1971; Skoog and Bjursell, 1974) and produces a varying dilution of the specific activity of [³H]dTTP arising from exogenous thymidine.

12.1.1. Overcoming the problem

There are three ways in which some of these problems may be overcome:

1. Flood the dTTP pool with exogenous [³H]dTTP.
2. Block the endogenous pathway.
3. Make allowance for the contribution of endogenous dTTP to DNA thymine.

12.1.1.1 Flooding the pool

By measuring the incorporation of thymidine into DNA from [³H]thymidine supplied at different concentrations to mouse L cells, Cleaver (1967) was able to show that at about 10^{-5} M thymidine incorporation reached a plateau, and a similar observation has been made for CHO cells (Fig. 12.5). This has been interpreted as showing that at this concentration the contribution of endogenous dTTP to DNA thymine is negligible. Care must be taken, however, that at higher thymidine concentrations inhibition of ribonucleotide reductase is not causing limitation in the supply of dCTP (see § 11.8.3). Experiments are better carried out at 5 or 10×10^{-5} M thymidine in the presence of 5–10 μM deoxycytidine.

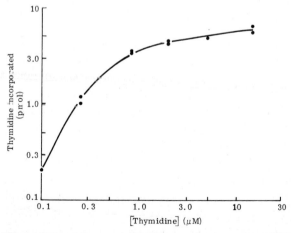

Fig. 12.5. Effect of thymidine concentration on incorporation. CHO cells were incubated with tritiated thymidine supplied at various concentrations and incorporation of radioactivity into DNA measured (for details, see legend to Fig. 12.6).

Example

- Set up 5 cm dishes containing 6×10^5 L929 cells in 5 ml of Eagle's MEM supplemented with 10% calf serum and buffered with Hepes and incubate overnight.
- Without allowing the cells to cool add 50 μl of a solution containing [³H]thymidine (50 μCi/μmol) and deoxycytidine so as to give final concentrations of 10^{-4} M thymidine (5 μCi/ml) and 10^{-5} M deoxycytidine. To maintain temperature the dishes should be kept in a humidified container inside a 37°C hot room.
- After 60 min stop incorporation by washing the cells in the dishes with:
  (a) cold BSS (twice)
  (b) cold 5% TCA (4 times)
  (c) absolute ethanol (twice)
- Air-dry and dissolve the cells in 1.0 ml 0.3 N NaOH (heat to 37°C if necessary).
- Rock the dish to mix and remove 0.5 ml to a counting vial.
- Neutralise with 0.1 ml 1.5 N HCl and add 6 ml of triton toluene scintillator (0.5% diphenyloxazole in a 2:1 mixture of toluene and triton X-100) and count.

12.1.1.2. Blocking the endogenous pathway

Siegers et al. (1974) showed that in the presence of amethopterin (10 μM), hypoxanthine (30 μM) and glycine (100 μM) DNA synthesis could be followed by measuring the incorporation of tritiated thymidine present at 3–30 μM. The rate of incorporation was linear and calculations based on the specific activity of the tritiated thymidine agreed with the amount of DNA made measured by other methods. Moreover, the presence of aminopterin did not affect the cell proliferation rate, the cell cycle time or the duration of S phase.

The increase with time in the incorporation of radioactivity into DNA is linear between 30 and 90 min of the concomitant addition of drug and tritiated thymidine and during this time the specific activity of the acid soluble nucleotide pool remains constant and identical to that of the supplied [³H]thymidine.

Example

- Set up cultures of hamster CHO cells in glass scintillation vials (2×10^5 cells in 1 ml of Eagle's MEM supplemented with 10% calf serum, hypoxanthine (30 μM) and glycine (100 μM), and buffered with Hepes buffer.
- After overnight incubation add 50 μl of a solution of amethopterin (2×10^{-5} M) and tritiated thymidine (2×10^{-4} M, 0.3 mCi/μmol).
- At 15 min intervals fix cells in the vials as follows:
  (a) wash twice with cold BSS (5 ml)
  (b) wash four times with cold TCA (5 ml)
  (c) wash twice with ethanol (5 ml)
  (d) wash with ether (2 ml) and air-dry.
- Dissolve the cells in 0.3 ml hyamine hydroxide (heat to 60°C for 10 min if necessary) and add 5 ml toluene scintillator (0.5% diphenyloxazole in toluene) and determine the level of radio-activity.
- Subtract the 30 min radioactivity count from the 90 min count or simply take the 60 min count if the incorporation is linear from zero time.

12.1.1.3. *Allowing for endogenous dTTP* (Adams, 1969b)

The problems with flooding the pool are the assumptions generally made that the concentration of thymidine used is (a) sufficient to flood the pool and (b) low enough not to have any deleterious effects on cell growth.

The rate of incorporation of radioactivity is dependent on the specific activity of the [³H]dTTP which in the two previous methods has been assumed to be the same as that of the supplied [³H]thymidine. As the concentration of [³H]thymidine is increased from low values (i.e. less than 10^{-6} M) the specific activity of the [³H]dTTP pool rises and so does the incorporation of radioactivity into DNA. This is at a time when the true rate of DNA synthesis remains unchanged. Thus the proportion of the dTTP pool which is radioactive, [³HdTTP]/([dTTP] + [³HdTTP]), is equal to the proportion of DNA thymine which is radioactive ([³H]thymine/total thymine).

This equation may be rearranged to give:

$$\frac{1}{[^3H]\text{thymine}} = \frac{[dTTP]}{\text{total thymine}} \cdot \frac{1}{[^3HdTTP]} + \frac{1}{\text{total thymine}}$$

which is the equation for a straight line cutting the ordinate at the reciprocal of the rate of DNA synthesis. Moreover, when the endogenous and exogenous pools of triphosphate (dTTP and [^{3}H]dTTP) are equal, the rate of incorporation of tritium is half maximal, which enables the concentration of the endogenous dTTP pool to be calculated. The assumption here is that the acid soluble deoxythymidylate pool is predominantly dTTP.

Example

- Set up coverslip cultures of mouse L929 cells in a 24 well tissue culture tray. Put 2×10^5 cells in each well in 0.5 ml Eagle's medium supplemented with 10% calf serum and incubate overnight.
- To each well add 10 μl of a solution containing 2 μCi of [^{3}H]thymidine at the following concentrations: 25 μM, 50 μM, 100 μM, 200 μM, 300 μM, 500 μM, 700 μM and 1 mM.
- Incubate for 60 min and then remove the coverslips and wash them by dipping them successively into three beakers of ice-cold BSS.
- Put the coverslips in a series of scintillation vials, each containing 0.5 ml cold 5% TCA and stand for 10 min.
- Remove the coverslips and wash by dipping successively into 4 beakers of 5% TCA and two of absolute ethanol.
- Put the coverslips in a second series of scintillation vials, add 0.3 ml hyamine hydroxide to dissolve the cells and count in 5 ml toluene scintillator to obtain a measure of the radioactivity in DNA ([^{3}H]thymine).
- To the cold TCA extract from stage 4 above add 5 ml of triton toluene scintillator to obtain a measure of the incorporation of ^{3}H radioactivity into the acid soluble pool ([^{3}H]dTTP).

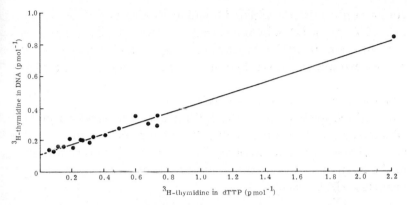

Fig. 12.6. Allowing for endogenous dTTP. 10^5 CHO cells in 0.5 ml Eagle's MEM–Glasgow Modification, supplemented with proline and 10% calf serum, were added to each well of a tissue culture tray. About half the cells attached to the coverslips in the wells and 30% of the cells could be shown to be making DNA by autoradiography, 10 μl of a solution containing 0.5 μCi [^{3}H]thymidine was added to each well to give a final thymidine concentration ranging from 0.1 to 15 μM. After 40 min incubation at 37°C the coverslips were removed and processed to estimate incorporation of radioactivity into DNA and acid-soluble material (dTTP). The reciprocals of these values are plotted against each other and the best straight line is drawn through the points. The correlation coefficient is 0.99. The intercept indicates a rate of thymidine incorporation into DNA of 8.9 pmol/40 min/5 × 10^4 cells or 0.97 μg DNA/h/10^6 S-phase cells. By measuring the tritiated acid-soluble pool size at half maximal incorporation the size of the endogenous dTTP pool is found to be 62.5 pmol/10^6 cells. Assuming the volume of 10^6 nuclei to be 0.1 μl this gives a concentration of 0.63 mM.

– Convert c.p.m. to d.p.m. and then to pmoles taking into account the varying specific activity of the tritiated thymidine.
– Plot [^{3}H]thymine^{-1} against [^{3}H]dTTP^{-1} and extrapolate to the ordinate to get total thymine^{-1} (Fig. 12.6). The value on the abscissa corresponding to 2 × total thymine is the amount of endogenously synthesised dTTP present in the cells which are making DNA.

Wittes and Kidwell (1973) have described a kinetic approach to measuring the pool size of dTTP which involves growing cells continuously in 10 μM tritiated thymidine which they found to label about 20% of the DNA thymine residues. They calculated that at this concentration 10^6 L929 cells growing in suspension would convert 9.5 pmoles extracellular thymidine into dTTP per min during S-phase.

12.1.1.4. Comparison of the methods

None of the methods is totally satisfactory. The first two involve addition of agents designed to alter the cell metabolism and the third is unsuitable for autoradiographic analysis. However, all three methods are attempts to overcome problems which many experimenters prefer to neglect.

12.1.2. Application to suspension cultures

When applied to suspension cells the acid washing steps may be done by repeated centrifugation but this is tedious and leads to losses of cellular material unless care is taken. Alternatives involve retaining the cells on glass fibre or cellulose acetate filters held in a micro-analysis filter holder (Millipore Corp. Ltd.); or the cells may be pelleted and then dissolved in a known small volume of 0.3 N NaOH. An aliquot is then allowed to soak into a Whatman 3 MM paper disc (2.5 cm diameter) and many discs may be processed together by washing in a beaker containing 5% TCA (4 times), ethanol (twice) and ether. The cellular material may be dissolved off the paper or glass fibre discs with hyamine hydroxide and counted in toluene scintillator.

12.2.   Estimation of rates of RNA and protein synthesis

[5-^{3}H]uridine is generally used as a precursor specific for RNA but this is only partially true as significant amounts of radioactivity may enter DNA particularly in the form of cytosine (Adams, 1968; Oldham, 1967). Many of the problems encountered with the use of [^{3}H]thymidine also arise in the use of [^{3}H]uridine but very little has been done in the search for a rigorous solution.

Radioactive amino acids are commonly used to follow protein synthesis. The procedure is in general similar to that used to follow synthesis of DNA, but in this case the cell's growth medium already contains high concentrations of amino acids and so the added radioactive amino acid acts as a tracer without causing imbalance. However, when it is desired to increase the extent of radioactive incorporation the presence of amino acids in the growth medium may be a disadvantage and their concentration may be reduced. Drastic reduction is obviously deleterious to cell growth (§ 11.7) and preliminary experiments must be performed to ascertain what reduction can be tolerated under the experimental conditions.

If a stationary culture of mouse L929 cells is subcultured in the absence of methionine the cells fail to enter S phase and even the presence of 10% the normal amount of methionine may have serious effects (Turnbull and Adams, 1975). However, once the cells have entered S-phase DNA synthesis will continue in the virtual absence of methionine, but the DNA and probably other polymers made are deficient in methyl groups.

Taylor and Stanners (1967) have shown that large polysomes begin to break down within 10 min of reducing the valine content of medium to 5% normal. This has the result of reducing the rate of protein synthesis measured using a variety of labelled amino acids. Thus although for the same outlay of isotopic amino acid the counts incorporated may rise when depleted medium is used, this may well reflect a reduced incorporation in molar terms.

12.3. Autoradiography

Because of the low energy of the β particles emitted on its decay tritium is ideally suited to high resolution autoradiography. The position of the silver grains is usually 0.1–0.5 μm from the source of the β particles for tritium, whereas it may be up to 290 μm for β particles from ^{14}C (Cleaver, 1967). The actual range depends on the energy of a particular β particle and the density of the material through which it must pass, and ^{14}C may be used if resolution is required only down to about 20 μm.

As well as giving high resolution the low energy of the tritium β particle poses a problem. As fixed cells are commonly at least 2 μm thick, only a proportion of the β particles will succeed in passing through the cell and reaching the autoradiographic emulsion. Cleaver and Holford (1965) estimated that because of self-absorption of β particles by the cells and also because at least half the β particles are emitted in a direction away from the emulsion, it takes 19 disintegrations of tritium to produce one grain in the emulsion. Those β particles which are emitted at an angle and have to travel through a thicker layer of biological material may never reach the emulsion. One advantage of the low energy of the β particles from tritium is that those particles reaching the higher density emulsion will be stopped within about 0.15–1 μm. Thus as long as the emulsion is thicker than 1 μm, its actual thickness will have little effect on sensitivity (Doniach and Pelc, 1950).

On the other hand, β particles from ^{14}C will travel, on average, 10 μm through emulsion, and so a thicker layer of emulsion while reducing resolution has the effect of increasing the number of grains produced. This has been made use of in double label techniques where DNA has been labelled with tritiated and ^{14}C-labelled thymidine and the cells covered with two layers of emulsion (Baserga, 1961; Dawson et al., 1962). The lower layer contains grains produced by both tritium and ^{14}C β particles but only the latter penetrate to form grains in the upper layer. By focussing separately on the two layers only those cells labelled with ^{14}C may be detected. This method has been used by Lala (1968) to measure the passage of cells in and out of S-phase.

12.3.1. Emulsions

The sensitivity and resolution of an autoradiograph depend on the emulsion used. Emulsions come in two forms: (a) as a gel which needs to be melted and diluted for use by a dipping procedure, and (b) as a preformed sheet which needs to be stripped from a glass plate, floated on a water surface and transferred to the specimen. Examples of the former are Ilford L4 and Kodak NTB3 and of the latter, Kodak AR10 (Appendix 3). The grain size of these emulsions varies from 0.12 μm (Ilford L4) up to about 0.4 μm (Kodak AR10) and each β particle hitting the emulsion produces from about 0.5 grains (Kodak films) to 1.3 grains (Ilford L4) (Caro, 1966). Although the use of stripping film was very popular, it is a more difficult and time-consuming procedure and offers no real benefit for tritium autoradiography. However, the even layer of film obtained is an advantage in quantitative autoradiography of ^{14}C-labelled compounds.

12.3.2. Stripping film

1. Coverslips covered with radioactively labelled cells are fixed, cells uppermost, to slides using DePex mounting medium (Gurr, Appendix 3). For best results the slides should be 'subbed', i.e. dipped into a 0.5% solution of gelatin containing 0.05% chrom alum and allowed to dry.

2. All further work must be carried out in the dark room using a red safe light.

3. Take a plate of stripping film (12.1 × 16.5 cm) and cut it with a razor blade. Cut 1 cm inside and parallel to each edge and discard the narrow strips. Cut the central portion into eight squares.

4. Within a few minutes of cutting the pieces curl up. They are then removed with scalpel and forceps and turned over onto the surface of clean distilled water at 20°C.

5. Leave the film for 3 min during which time it swells and expands.

6. Immerse the slide bearing the cells below the water surface and withdraw it so that the film drapes itself about the slide. The long edges of the film fold round underneath the slide.

7. Drain and dry with a stream of cold air.

8. Expose and develop as indicated below.

One major difficulty with stripping film is that the film may move relative to the specimen during processing. This is usually avoided if the swelling (stage 5) and drying processes are adequate.

12.3.3. Liquid emulsion

1. Fix the cell bearing coverslips to slides as above. The use of subbed slides is unnecessary.
2. In the dark room thaw the emulsion by standing the bottle in a water bath at 40°C. As this is a lengthy process and one which it is best not to repeat too often on one batch of emulsion, it is preferable to scoop out 5–10 g quantities of the solid gel into universal containers, which should be stored separately. When required one container may be thawed.
3. Dilute the liquid emulsion with one or two volumes of distilled water. Avoid producing bubbles by shaking or vigorous stirring.
4. Dip the slides into the diluted emulsion. If the coverslip is attached to one end of the slide the depth of liquid emulsion need be no more than 2–3 cm.
5. Drain off excess emulsion and dry in a stream of cold air. Horizontal drying produces a thicker layer of emulsion. If a more dilute emulsion is used then drying in a vertical position may produce a layer of emulsion less than 1 μm thick.
6. Autoradiographs are best exposed in a light tight box for 3–7 days. Exposure for longer periods is reported to cause fading of the latent image, unless the autoradiographs are thoroughly dry and preferably stored in an oxygen free atmosphere at low temperature. This may be achieved by placing a small bag of desiccant in the light tight box along with a lump of solid CO_2. The whole may be sealed with tape and stored in a refrigerator. N.B. ensure that no high energy radioactive source, e.g. ^{32}P, is also in the refrigerator. However, if the layer of emulsion is more than 1 μm thick the grain counts will increase in a linear fashion up to 9 days even when the slides are stored in air at room temperature (Fig. 12.7).
7. After exposure immerse the slides in Kodak D19b developer at 20°C for 3–5 min.
   This may be prepared by dissolving the following in the order given:

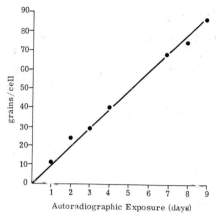

Fig. 12.7. Effect of length of autoradiographic exposure on grain count. BHK21/C13 cells labelled with [³H]uridine were covered with Ilford L4 emulsion (diluted with an equal volume of water), dried in a horizontal position and exposed in air at room temperature for the indicated times. (Courtesy of K. Shaw and J.D. Pitts.)

2.2 g metol (Ilford Ltd., Appendix 3)
144 g hydroquinone
48 g anhydrous sodium carbonate
4.0 g potassium bromide.
Make up to 1 l with distilled water and store at 4°C in FULL bottles (to prevent oxidation). Ensure that the developer is warmed to 20°C before use.

8. Fix in Amfix (May and Baker, Appendix 3) diluted with two parts of water for 5 min or twice the clearing time and rinse in tap water.

9. Stain with Giemsa for 3 min and wash thoroughly with tap water. Air-dry.

10. Fix a second clean coverslip onto the specimen using DePex. Unless the labelling is heavy it is usually necessary to use oil immersion optics.

12.3.4. Autoradiography in dishes

Cells growing directly in dishes (5 cm Petri dishes) may be labelled with a radioactive precursor and fixed in a similar way to cells on coverslips. One ml of diluted liquid emulsion is then added and the material exposed (without the lid) and developed as indicated above.

12.3.5. The value of grain counting

Autoradiographic analysis is the only method which can determine the proportion of cells incorporating a radioactive precursor and the site of that incorporation. Thus, tritiated thymidine is incorporated into DNA in the nuclei of those cells in S-phase and tritiated hypoxanthine appears first in the nucleus and later in the cytoplasm of cells with HPRT but not in mutants lacking the enzyme (see § 13.1).

Grain counting, however, as a quantitative measure of the rate or extent of a given process, such as DNA synthesis, suffers from many of the same disadvantages as does measurement of the total incorporation of radioactivity by a culture (see § 12.1). Thus [³H]thymidine is diluted by endogenous pools of dTTP which may differ in extent among cells. In addition, the lack of an accurate figure relating grain count to disintegrations means that at best grain counts are a relative measure of rates of synthesis of macromolecules. However, grain counting is the only way in which rates of incorporation by individual cells may be compared.

Incubation of cells with [³H]thymidine not only reveals that only a proportion of cells are making DNA. It also shows that the number of grains per labelled cell is very variable – far more so than would be expected on statistical grounds. Figure 12.8 shows a distribution of grain counts compared with a Poisson distribution with a mean of 30 grains per cell. The wide distribution is caused partly by cells entering or leaving S-phase during the period of labelling and this may be reduced by shortening the pulse time so that very few cells fall into this category. Another cause of the wide distribution of grain counts is the fact that the rate of DNA synthesis is not constant over S-phase (§ 10.3) and the population studied will have cells at all stages of DNA synthesis.

12.3.6. Background grains

Apparent in Figure 12.8 is the group of 'unlabelled' cells. Such cells usually average less than one grain per cell and a similar distribution of grains is found over regions of the slide where there are no cells. It is common practice to count the grains over such a control area to get a measure of true background.

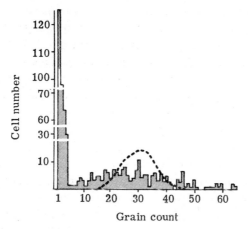

Fig. 12.8. Grain count distribution. L cells labelled for 10 min with [³H]thymidine (2.5 μCi/ml; 0.36 Ci/mmol) and processed for autoradiography using NTB3 emulsion. Those cells (62% of the total) with 1 or more grains are recorded. A Poisson distribution with a mean of 30 is included for comparison. (Reproduced from Cleaver, 1967, with kind permission of the author.)

Higher background are found: (1) if the emulsion used is old or has been stored adjacent to a radioactive source, e.g. ³²P; (2) if exposure times are long; or (3) if removal of tritiated precursors or acid soluble components has been inadequate.

12.3.7. Autoradiography of water-soluble cell components

In the previously described methods the radioactive precursor molecules and acid soluble intermediates have been rigorously removed to reduce background grains so that incorporation has been measured only into macromolecules. However, a number of procedures have been developed to visualise the localisation of water soluble compounds within cells. As an initial step these methods rely on the efficient removal of extracellular radioactive material by several washes in saline. It is not clear whether such washes also remove intracellular material which is able to diffuse out of the cells under the washing conditions. Thus little tritiated thymidine remains associated with cells washed three times with ice-cold Earle's BSS but this could mean either (1) there is no intracellular thymidine or (2) intracellular thymidine rapidly diffuses out of cells washed with BSS. Thus the autoradiographic procedures described below will detect charged molecules which are generally extracted from cells only with acid fixatives, e.g. thymidine phosphates, glucose phosphate, etc.

12.3.7.1. Cell fixation

The method used depends largely on whether precise localisation is required or whether a certain amount of diffusion of the soluble compounds can be tolerated or is required. In the latter case rapid air drying (at 37°C) of cells labelled with tritiated thymidine allows thymidine phosphates to diffuse a little way to form a halo around the nucleus (Adams, 1969a). For precise localisation it is necessary to freeze the cells in isopentane or freon held at liquid nitrogen temperature and then to subject them to lyophilisation. This drying process is very rapid as the cells are so thin and may be done by placing the slides on a metal block precooled in liquid nitrogen and placing the block in a desiccator attached to a vacuum pump. Alternatively, the slides may be placed in a tube held in a salt/ice mixture at $-20°C$ and the tube connected to a vacuum pump. In both cases traps containing methanol/dry ice must be present to trap the water vapour.

Rather than being lyophilised cells may be *freeze substituted* (Pearse, 1953). After treating cells with isopentane at liquid nitrogen temperature they are flooded in several changes of absolute methanol at the temperature of solid CO_2 for 2–3 h.

12.3.7.2. Covering with emulsion

Once again if a certain amount of diffusion can be tolerated the fixed slides may be dipped into liquid emulsion and quickly dried in a horizontal position (Adams, 1969a). Drying vertically leads to a stream of grains trailing away from cells as the soluble radioactive compounds are washed out by the liquid emulsion

Fitzgerald et al. (1961) cooled fixed slides in a refrigerator so that when they were brought out into the warm dark room a layer of condensation formed on the surface. This is sufficient to allow a square of dry stripping film to be stuck to the slide by means of thumb pressure. Although the film sticks at this stage it usually swells on developing, leading to considerable movement relative to the cells.

To avoid the problems of unswollen stripping film and to reduce diffusion to a minimum, partly dried liquid emulsion can be added to fixed slides (Miller et al., 1964a; Finbow and Pitts, 1979).

-- Place a small amount of diluted liquid emulsion in a 9 cm Petri dish at 44°C and allow it to cool until it begins to set.

– Dip a wire loop of about 8 cm diameter into the emulsion and withdraw it covered in emulsion.

Place the loop over the slide lying on a horizontal surface. The emulsion should form a neat circle round the slide.

– Dry using a cold fan and expose as for normal autoradiography.

12.4. DNA repair

Incorporation of tritiated thymidine into DNA as a result of repair seldom interferes with studies of DNA synthesis. However, following irradiation, or in the presence of certain drugs which suppress DNA replication, the major reason for incorporation of tritiated thymidine may be to help repair DNA.

DNA repair is best studied in a system where the background levels of replication are low. Suitable systems are cultures which have come to rest at high density, or unstimulated lymphocyte preparations where less than 1% of the cells is in S-phase. The low levels of replicative incorporation can be further repressed by 1–2 mM hydroxyurea which selectively inhibits replication (Cleaver, 1969b). This selective effect may simply be a result of the very small pools of deoxyribonucleoside triphosphates required for repair.

That incorporation occurring under these conditions is a result of DNA repair can be confirmed by labelling with [³H]bromodeoxyuridine in the presence of 1 μM fluorodeoxyuridine and hydroxyurea. The DNA is then isolated and centrifuged to equilibrium on a gradient of caesium chloride when the label is found in the light, unsubstituted position (Cleaver, 1969a; Abo-Darub, 1977).

12.4.1. Ultraviolet irradiation

For irradiation cells must be present as a monolayer in a plastic (UV transparent) dish or bottle and must be immersed in a minimum volume of PBS-A or some other liquid which does not absorb UV light. They are then exposed to light of 254 nm emitted at a dose of about 0.5 J.m^{-2} s^{-1} at a suitable distance (Abo-Darub et al., 1978). The energy emission of the lamp may be measured using a chemical actinometer (Hatchand and Parker, 1956) which relies on the fact that irradiation of a solution of potassium ferrioxalate releases ferrous ions in an amount proportional to the dose of ultraviolet radiation.

After irradiation growth medium is added to the cultures which are quickly returned to the incubator.

12.4.2. Estimation of repair synthesis

Label cells with [³H]thymidine (5 μCi/ml, 20 Ci/mmol) in the presence of 1–2 mM hydroxyurea. After the labelling period harvest the cells in the normal way for measuring incorporation into DNA. Alternatively, cells may be processed for autoradiography when the presence of hydroxyurea is not necessary as S-phase cells are readily recognised by their very high grain count. Unirradiated cells should show less than 0.2 grains per cell while those irradiated at 5 J.m^{-2} should show 15–20 grains per cell after exposure for 8 days (Abo-

Darub et al., 1978). Figure 12.9 shows the time course of repair for irradiated lymphocytes from normal subjects and those with actinic keratosis (a disease, prevalent in certain regions, which produces tumours on areas of skin exposed to sunlight). The rate of DNA repair is slower than normal in the lymphocytes from patients with actinic keratosis.

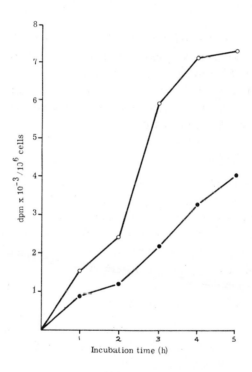

Fig. 12.9. Time course of [³H]thymidine incorporation into DNA of UV irradiated, hydroxyurea-treated lymphocytes. 3×10^6 lymphocytes from actinic keratosis patients (●) or age-matched normal individuals (○) were incubated with [³H]thymidine (5 μCi/ml, 18.5 Ci/mmol) and hydroxyurea (1.5 mM) at 37°C immediately after irradiation (20 J.m^{-2}). After incubating for different times cells were fixed in Carnoy's fixative and washed with 5% trichloracetic acid and ethanol before counting. (Reproduced from Abo-Darub et al., 1978, with kind permission of the authors and publisher.)

CHAPTER 13

Cell mutants and cell hybrids

Major advantages that bacteria hold for biochemists are the ready availability of mutants coupled with a short generation time. Compared with whole animals, animal cells in culture have a relatively short generation time and so a major effort has gone into the production and selection of mutants of animals cells and their use in the study of somatic cell genetics.

13.1. Auxotrophic mutants

Kao and Puck (1968) have developed a general method for production of auxotrophic mutants. This is based on the observation that DNA containing 5-bromodeoxyuridine (BUdr) is sensitive to visible light produced by fluorescent lamps. This can be used to selectively kill prototrophs growing on restrictive media, i.e. media which restrict the growth of auxotrophic (or other) mutants. The cells that later grow out in supplemented media are auxotrophic mutants. The successive steps involved are given below.

- Grow cells on enriched media containing the mutagen N-methyl-N'-nitro-N-nitrosoguanidine (MNNG) at 0.5 μg/ml or ethyl methane sulphonate (EMS) at 200 μg/ml for 16 h.
- Transfer the cells to mutagen-free enriched medium for several generations to allow the growth of both prototrophs and mutant auxotrophs.
- Transfer the cells to growth on minimal restrictive medium for sufficient time for the pools of essential metabolites to become depleted.

– Add BUdr (10^{-6} M) to the culture. The BUdr is incorporated into the DNA of the prototrophs which are the only cells growing in minimal restrictive medium.

– After 2 cell generations remove the BUdr and expose the cells for 30 min to light from a 40 watt 'Cool White' fluorescent lamp (General Electric). Prototrophs are killed.

– Transfer to complete enriched growth medium when clones of auxotrophs appear after 1–2 weeks.

13.2.   Selection of mutants

Mutants defective in particular enzymes can be selected if two conditions are fulfilled: either the enzyme is not essential or an analogue of the normal substrate leads to lethal incorporation.

Cells lacking thymidine kinase (TK⁻ cells) can be isolated by treating cell cultures with high concentrations (30 μg/ml) of 5-bromodeoxyuridine, which kills cells containing the enzyme thymidine kinase due to incorporation of large amounts of the analogue into the cells' DNA

$$5BUdr \rightarrow 5BdUMP \rightarrow 5BdUDP \rightarrow 5BdUTP \rightarrow DNA$$
thymidine
kinase

Mutants lacking, or with much reduced levels of the enzyme thymidine kinase survive as this enzyme is not essential. Thus the cell can make dTMP from dUMP using folic acid as the one carbon donor (Fig. 13.1).

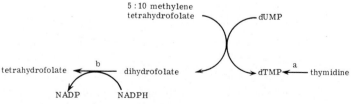

Fig. 13.1. Synthesis of dTMP. a = Thymidine kinase; b = dihydrofolate reductase.

In a similar way mutants lacking hypoxanthine phosphoribosyl transferase (HPRT) can be isolated by growing cells in 8-azaguanine (3 μg/ml). This analogue of guanine is incorporated into the purine nucleotide pool and hence into nucleic acids by the phosphoribotransferase which normally would be involved in the uptake of hypoxanthine or guanine. Once again only mutant cells lacking the enzyme survive. The enzyme is not essential as IMP is synthesised endogenously from smaller precursors. Two steps in the endogenous pathway utilise folic acid to add single carbon units to the growing purine ring (Fig. 13.2).

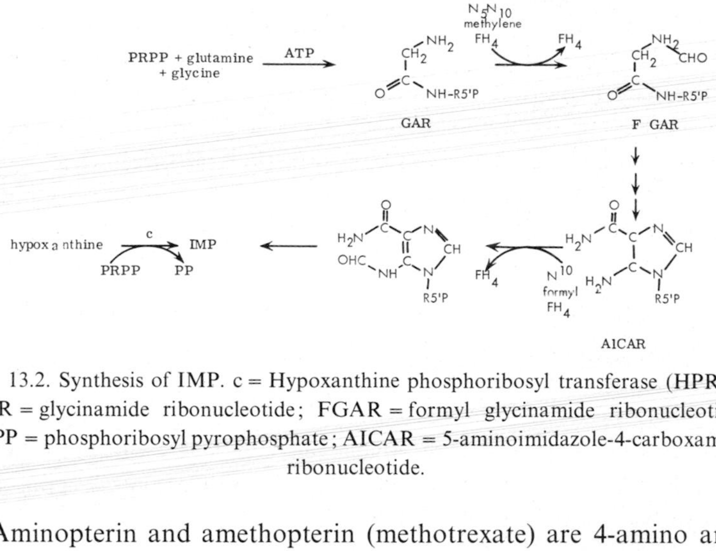

Fig. 13.2. Synthesis of IMP. c = Hypoxanthine phosphoribosyl transferase (HPRT); GAR = glycinamide ribonucleotide; FGAR = formyl glycinamide ribonucleotide; PRPP = phosphoribosyl pyrophosphate; AICAR = 5-aminoimidazole-4-carboxamide ribonucleotide.

Aminopterin and amethopterin (methotrexate) are 4-amino analogues of folic acid (Fig. 13.3) and interfere with the production of the active intermediates by blocking the enzyme dihydrofolate reductase (reaction b) and so breaking the cycle (Fig. 13.4).

The action of the antifolate drugs would therefore be expected to lead to accumulation of dihydrofolate, but this was found not to be the case by McBurney and Whitmore (1975) who suggested that the drugs may also have a direct inhibitory action both on the folate requiring enzymes involved in purine biosynthesis and on thymidylate

Folic Acid

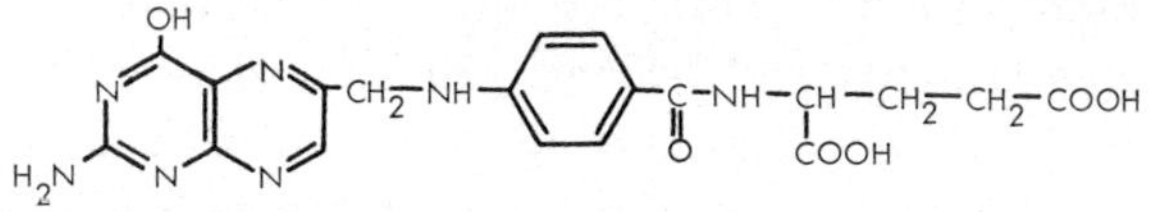

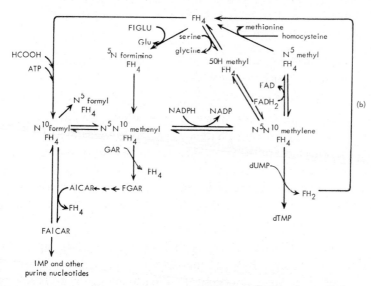

Aminopterin (R = H) and Amethopterin (R = CH$_3$)

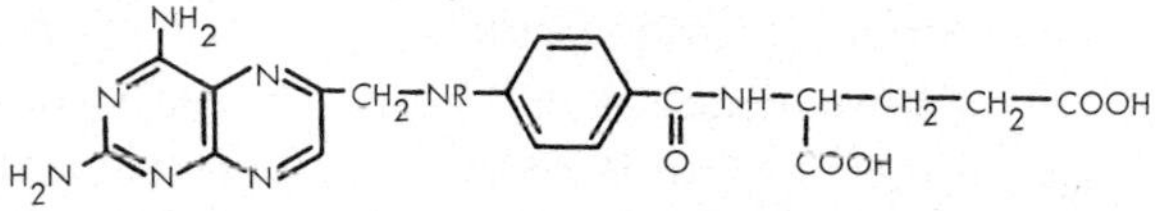

Fig. 13.3. Structure of folic acid and its analogues.

Fig. 13.4. Interconversions of tetrahydrofolate derivatives. FH$_2$ = dihydrofolic acid; FH$_4$ = tetrahydrofolic acid; AICAR = 5-aminoimidazole-4-carboxamide ribonucleotide; FAICAR = formyl AICAR; GAR = glycinamide ribonucleotide; FGAR = formyl GAR; Glu = glutamic acid; FIGLU = formimino glutamic acid. (Modified from Mudd and Cantoni, 1964.)

synthetase. However, by growing cells in the presence of increasing concentrations of aminopterin a number of resistant cell lines have been isolated (Hakala and Ishihara, 1962; Littlefield, 1969). These have been characterised as having either an altered permeability or an altered folate reductase or an increased rate of synthesis and hence increased amounts of the enzyme (Alt et al., 1976), resulting, at least in part, from a selective amplification of the dihydrofolate reductase gene (Alt et al., 1978; Schimke et al., 1978). The problem is considered in more detail in § 11.8.1. The importance of the antifolates lies in their role in the HAT selection technique (§ 13.5) devised by Littlefield (1964) for the isolation of hybrids between mutant cells defective on the one hand in thymidine kinase and on the other hand in HPRT (see below). They are also used to suppress the endogenous pathway when labelling cells with radioactive thymidine or hypoxanthine (§ 12.1.1.2), to synchronise cells by depleting the pools of thymidine (§ 11.8.1), and in the clinical field as tools in cancer chemotherapy.

13.2.1. *Procedure for isolation of TK⁻ mutants* (Littlefield and Basilico, 1966)

– Seed 5×10^5 BHK21/C13 cells into 100 mm dishes in Dulbecco's modification of Eagle's medium supplemented with 10% calf serum, 10% tryptose phosphate and bromodeoxyuridine at 3.3 µg/ml. This kills most of the cells and after 10 days only about 20 cells have survived to form colonies.
– Harvest the cells and reseed in the same manner but increase the bromodeoxyuridine concentration to 30 µg/ml. This selects for highly resistant mutants.

After 10 days a further 20 or so clones will have grown up which are found to have only 1–3% of the thymidine kinase activity of the parent cells and which incorporate [¹⁴C]thymidine into DNA at 5–7% the rate of the parent cells.

13.3. Temperature sensitive mutants

The general method of isolating a temperature sensitive mutant involves exposing a mutagenised cell population to an agent lethal to dividing cells at the non-permissive temperature. That temperature is usually 39°C, but 37.5°C has been found more useful in reducing the number of leaky mutants (i.e. mutants which show a high tendency to revert to normal) obtained (Basilico, 1977, 1978; Talavera and Basilico, 1977).

Lethal agents include compounds which lead to unbalanced growth by blocking DNA synthesis, e.g. fluorodeoxyuridine or cytosine arabinoside, or agents which are incorporated into DNA where they have deleterious effects, e.g. high levels of tritiated thymidine or bromodeoxyuridine which can lead to breaks in the DNA on subsequent irradiation (Thompson et al., 1970).

In general the mutants isolated are temperature sensitive for DNA synthesis (Sheinin, 1976; Nishimoto et al., 1978), but Basilico (1978) indicates the following procedures for selecting for G1 and G2 ts mutants.

1. Selection of G1 mutants and S mutants:

a. mutagenise the cells as described in § 13.1,

b. synchronise the cells in G1 by isoleucine deprivation (§ 11.7),

c. release at 39°C (see § 11.7) in the presence of 5 μM 5-fluorodeoxyuridine for 2 days, or

d. release at 33°C in the presence of 2 mM hydroxyurea and after 10 h change for medium at 39 °C minus hydroxyurea but plus fluorodeoxyuridine.

Incubate for 12–16 h.

Point c will yield G1 mutants as mutants in other phases will grow in an unbalanced fashion and die. Point d yields S-phase mutants for a similar reason.

2. Selection of G2 mutants and M mutants:

a. mutagenise the cells as described in § 13.1,

b. shift to 39°C for a generation time when all ts mutants will be blocked,

c. detach mitotic cells (§ 11.2) which will include ts mutants blocked in mitosis,

d. incubate at 39°C for a generation time in the presence of fluoro-deoxyuridine which by causing unbalanced growth leads to the death of wild type cells (§ 11.8),

e. reduce the temperature to 33°C and select mitotic cells (§ 11.2) over the next hour or two; these will be G2 mutants.

In all cases the cells must be cloned (§ 8.1) and checked to ensure the selection process has worked satisfactorily.

13.4. Replica plating of animal cells

One of the tools which has enabled the bacteriologist to screen thousands of mutants is the ability to transfer large numbers of clones from one dish to another while retaining their orientation. In this way a thousand or more clones can be tested simultaneously for several growth characteristics and those of interest picked out from the original dish.

With animal cells the problems are that the cells grow firmly attached to the substratum yet at the same time move about so that a single clone very soon grows to cover a wide area. A method described recently by Esko and Raetz (1978) claims to circumvent these problems.

– Cell monolayers are treated with the mutagenic agent ethane methane sulphonate (400 μg/ml: Eastman Ltd.) and allowed to grow for 3 days.
– About 1000 cells are plated into a 10 cm dish containing 15 ml growth medium and incubated for 1 day.
– Float a disc of sterile Whatman No. 50 paper on the medium and weight it down with glass beads to form a single even layer.
– Change the medium every 2–3 days.
– After 7–10 days remove the medium; decant the beads and remove the paper disc to which over 95% of the cells are attached.
– Wash the paper disc in a stream of medium (30 ml) to remove any small clumps of cells.

- Float the paper disc (cells down) on a fresh dish of medium, weight down and incubate.
- Repeat the last three stages every 3 days when each new dish is a replica of the original.

The cells on the paper disc are viable and may be used in auto-radiographic studies. Alternatively, they may be lysed by freeze–thawing and incubated in enzyme assays.

An alternative method of achieving the same end is to plate out the mutagenised cells in a 96-well microtitration tray as for cell cloning. When colonies have grown up they can be harvested and distributed to several more trays using manual or automatic cell harvesters and diluters available from Titertek (Flow Labs. Ltd.).

13.5. Somatic cell hybridisation

This is a technique pivotal to the genetic analysis of cultured animal cells and to the results of which the journal 'Somatic Cell Genetics' is devoted. A book on cell hybrids was published in 1976 (Ringertz and Savage). Although such studies are crude relative to the genetic analysis in bacteria, it is now easy to locate genes to particular chromosomes. Traditionally, such studies have been performed by analysis of the offspring of parents showing particular phenotypic characteristics, but the process was taken to the level of biochemical characteristics following the observation that under particular conditions two cells in vitro will fuse to yield heterokaryotes, i.e. single cells containing two distinct nuclei. A small proportion of these will multiply indefinitely. The first mitotic division after fusion leads to daughter cells with both sets of chromosomes in the same nucleus. Very often at subsequent divisions chromosomes are lost in ones and twos, probably as a result of frequent mitotic abnormalities until a new stable cell line is formed carrying some chromosomes from each parent (Weiss and Green, 1967; Ruddle, 1973; Giles and Ruddle, 1973). In human/mouse and Chinese hamster/mouse hybrids it is always the mouse chromosomes which are retained (Matsuya and Green, 1969).

In the study of cell hybrids it is essential to be able (1) to select for heterokaryotes and (2) to recognise them. The simplest complementation analysis involves the fusion of cells differing in a single gene mutation. Thus if cells requiring glycine for growth (gly$^-$) are fused with cells which require hypoxanthine for growth (hyp$^-$) then, if the mutations are recessive, the heterokaryons will be able to grow in medium lacking both glycine and hypoxanthine. This provides a selection technique for the fusion products.

Similarly, if a mouse cell lacking the enzyme thymidine kinase (cannot grow in the presence of aminopterin + thymidine) is fused with a human cell which lacks the enzyme hypoxanthine phosphoribosyl transferase (HPRT) (cannot grow in the presence of aminopterin + hypoxanthine) then only the fusion products will grow in medium containing hypoxanthine (10^{-4} M), aminopterin (4×10^7 M) and thymidine (10^{-5} M) (HAT medium) (Littlefield and Goldstein, 1970). As stated above, such heterokaryons lose human chromosomes preferentially. However, one chromosome they cannot lose while still retaining the ability to grow on HAT medium is the one carrying the gene for thymidine kinase. By taking a number of clones derived from such a fusion event and comparing their human chromosome complement it soon becomes clear which is the essential chromosome and hence the localisation of the gene for the enzyme thymidine kinase could be established (Weiss and Green, 1967).

In this way it has been shown that HPRT is located on the X-chromosome and it is from this observation that much recent work has sprung. Attempts are now being made to fuse cells, the chromosomes of one of which have been fragmented by irradiation. The extent of cotransfer of a particular characteristic with HPRT into a stable clone gives a measure of the distance between the two markers and so some relatively fine genetic mapping can now be done using animal cell hybrids (Goss and Harris, 1975).

A related technique involves the isolation and fractionation of chromosomes from a donor cell followed by their uptake in fragmented form into a mutant recipient (Willecke et al., 1976a, b). Once again the extent of cotransfer of two particular genes gives a measure of the distance between them.

Other selection systems involve the fusion of normal human lymphocytes or ascites tumour cells (which grow poorly in vitro) and defective mouse cells (again in HAT medium) and the fusion of proline requiring CHO cells and defective mouse cells in HAT medium lacking proline.

Human cells are killed by 10^{-7} M ouabain, but mouse cells are relatively resistant in that 10^{-3} M is a lethal dose. Mouse human hybrids show intermediate sensitivity and hence HAT medium containing 10^{-7} M ouabain can be used to select for hybrids between any human cell and TK^- or $HPRT^-$ mouse cells. This enables primary human cells and other non-selected strains to be used in fusion experiments (Mayhew, 1972; Thompson and Baker, 1973).

13.5.1. Cell fusion with Sendai virus

Although cells sometimes fuse spontaneously, the frequency of fusion can be dramatically increased by treatment with a number of agents, the most commonly used of which is inactivated Sendai virus.

Virus which has been inactivated by UV irradiation or by treatment with β-propiolactone (Neff and Enders, 1968) is used (see § 14.2.4).

Cell fusion can be done with the cells growing as a monolayer or preferably in suspension. Sendai virus is difficult to control and the fusion process in monolayers can get out of hand resulting in formation of huge polykaryocytes containing tens of nuclei, most of which are non-viable. This can be limited by having the two cell types to be fused present in very unequal numbers (e.g. 1–1000). The mixed cell monolayer is allowed to grow for 24 h and then, after removal of the medium, the cells are treated with inactivated Sendai virus for 10 min at 0–4°C (when adsorption of the virus occurs), rinsed several times with serum-free medium and then returned to growth medium at 37°C. Under these conditions about 4% of the cells fuse, but the precise number depends largely on the cells being used. On returning the cells to 37°C the cell membrane adjacent to the site of adsorption of the virus becomes damaged and cytoplasmic bridges form between adjacent cells within 1 min. As the damaged sites are repaired the bridges become wider and giant cells form (Hosaka and Koshi, 1968). After 1 h at 37°C the cells should be harvested and reseeded at about 10^5/ml in tissue culture dishes or bottles.

The *standard procedure* for cell fusion with inactivated Sendai virus is as follows:

- The two cell suspensions to be fused are mixed in a screw capped tube. 5–10 × 10^6 cells of each type should be present in a total volume of 1 ml BSS.
- An equal volume (1 ml) of inactivated Sendai virus, 4 × 10^{10} particles or 1000 haemagglutination units (HAU)/ml (§ 14.3.3) is added and the pH adjusted to 8.0–8.5.
- Some procedures recommend shaking at 0–4°C for 30 min at this stage. Whether or not this is done, the tube should now be shaken gently at 37°C for 30–40 min. The cells clump immediately and the pH falls slightly.
- Dispense 1 ml suspension into a 5 cm dish with 4 ml complete medium (Watkins, 1971; Stadler and Adelberg, 1972).

13.5.2. Cell fusion of BHK21/C13 cells with chick red blood cells using Sendai virus

Avian red cells are nucleated, but Sendai virus, as well as causing agglutination, haemolyses the red cells so that fusion involves a red cell ghost containing the nucleus but little cytoplasm.

- Trypsinise (§ 5.2) the hamster cells and resuspend in Eagle's medium plus 10% calf serum. Wash twice by centrifuging and resuspending in serum-free Eagle's MEM (Appendix 1) (2 × 10^6 cells/ml).
- Harvest the chick red cells from a 17-day-old fertile egg as follows: remove the shell over the air sac and cut the blood vessels of the membranes allowing them to bleed into the allantoic fluid. Remove this and wash the cells in serum-free Eagle's MEM (0.5 ml).

- Mix 0.1 ml hamster cell suspension with 0.1 ml red cell suspension and 0.1 ml of inactivated Sendai virus (200 HAU) in a 15 ml sterile, centrifuge tube at 37°C for 25 min.
- Make the contents up to 5 ml by adding Eagle's MEM plus 10% calf serum and transfer to a 5 cm Petri dish (which should contain coverslips). Incubate at 37°C in a CO_2 incubator.
- Remove coverslips and wash twice in PBSA (Appendix 1) and fix in methanol (10 min). Stain for 3 min in May–Grunwald Giemsa stain (Appendix 2), wash in distilled water and acetone and mount on slides, cells down.

13.5.3. Cell fusion with lysolecithin

Because of the difficulties of using inactivated Sendai virus other methods of cell fusion have been developed of which the two simplest are fusion using lysolecithin or polyethyleneglycol.

- Co-cultivate overnight about 2×10^5 cells of each type in 4 ml medium.
- Remove the medium and wash the cell sheet twice with 4 ml of acetate buffer (0.1 M NaCl; 0.05 M NaAc pH 5.8).
- Add 4 ml of acetate buffer containing 100 μg lysolecithin/ml.
- After 15 min at 37°C remove the lysolecithin and wash the cell sheet twice with complete medium and finally return to the incubator with 4 ml complete medium (Croce et al., 1971; Keay et al., 1972).

13.5.4. Cell fusion using polyethyleneglycol

This method devised by Pontecorvo (Davidson et al., 1976; Pontecorvo et al., 1977) is the simplest method and the one coming into most frequent use.

- Prepare a mixed suspension of the cells to be used (10^7–10^8 of each) and sediment. This may be the most important part of the procedure but the efficiency is increased by the subsequent treatments.

- Resuspend the mixed pellet of cells in 0.5 ml 40–50% polyethylene-glycol (PEG 1500) in serum free medium (the presence of 15% dimethylsulphoxide may improve the yield of hybrid cells).
- After 1 min at 37°C add a further 0.5 ml 20% polyethyleneglycol in serum free medium.
- After 3 min dilute the suspension with, and wash the cells in complete medium before plating out in multiwell dishes in selection medium.

13.6. Cell communication

Although neither TK$^-$ cells nor HPRT$^-$ cells will grow separately in HAT medium, it is found that mixed populations of these two mutant lines will grow in HAT medium (see Fig. 13.5).

This metabolic cooperation is not a result of the selection of hybrid cells as it occurs almost instantaneously on mixing in the absence of any fusion agent. It is caused by the exchange of low molecular weight compounds through gap junctions. Thus the TK$^-$ cells obtain thymidylate from their HPRT$^-$ neighbours which get purine nucleotides in exchange. Other low molecular weight compounds which have been shown to pass gap junctions are tetrahydrofolate (Finbow and Pitts, 1979), and dyes of various sizes up to a molecular weight of 1000 (Pitts and Finbow, 1977; Bennett, 1973) which suggests that most small molecules are exchangeable between neighbouring cells.

For cell communication to occur it is essential that the cells are in contact. Using autoradiography the diffusion of nucleotides is readily apparent only between cells in contact, whereas cells equidistant from a neighbour yet not in actual contact do not communicate.

Cell communication may interfere with the selection of hybrids unless the cells are plated at sufficiently low densities that contact is only made occasionally. Alternatively, hybrids can be made using a cell line which does not communicate, and the L929 cell line is one

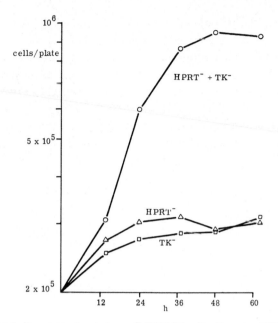

Fig. 13.5. Metabolic cooperation between BHK21/13 cells. Growth of BHK–HPRT⁻ and BHK–TK⁻ cells separately and in mixed (1:1) culture, in medium containing hypoxanthine, aminopterin and thymidine (HAT medium). HPRT = hypoxanthine phosphoribosyl transferase; TK = thymidine kinase. (Reproduced from Pitts, 1971, with kind permission of the author and publisher.)

which has lost this ability together with its HPRT⁻ sub-line (A9) (Pitts, 1971).

13.6.1. Grain counting and cell communication

Incubation of a population of cells with tritiated uridine or hypoxanthine leads to incorporation of radioactivity into the RNA of all but the mitotic cells. Analysis of the grain counts shows these to fall into a Poisson distribution, the median point of which corresponds to the relative rate of RNA synthesis of the culture. Incubation of HPRT⁻ cells with tritiated hypoxanthine leads only to a low background of grains.

   Mixed populations of cells, some of which lack HPRT, are readily distinguished when plated at low cell density. However, at higher cell densities where cell communication is possible the HPRT positive cells contribute radioactive nucleotides to their mutant neighbours through gap junctions and the distinction between the two cell types becomes less clear (Pitts and Simms, 1977). It may be resolved using a computer programme which not only reveals the presence of two cell types but also gives the mean grain count in each type.

CHAPTER 14

Viruses

14.1. Introduction

Cell culture is the predominant and indispensible tool for virus isolation and cultivation. Although some viruses are more easily isolated in animals and embryonated eggs, the modern era of virology only began when Enders et al. (1949) showed that poliovirus was able to reproduce in various kinds of human embryonic cells in culture whereas in vivo its multiplication is largely restricted to the neurons in the grey column of the spinal cord.

Primary cells, cell strains and established cell lines are now used for the isolation and propagation of animal viruses. In general, the procedure is the same for all virus : cell combinations.

Remove the medium from a cell monolayer and wash the monolayer with BSS or PBS (Appendix 1) to remove inhibitors (antibodies) which may be present in the medium. Apply the virus preparation suspended in a small volume of BSS or PBS and allow 30–60 min for adsorption. Replace or supplement the salt solution with fresh medium.

When viruses infect cultured cells they produce characteristic morphological changes. The end result of cellular degenerative processes (the cytopathic effect or CPE) may only be obvious after a week's growth in the presence of the virus but in other cases is obvious after as little as 12 h. The details of the morphological changes vary for different viruses.

If, rather than a productive infection, the virus brings about cellular transformation this also produces characteristic changes in cell morphology and growth characteristics. This is discussed in more detail in § 14.4.

A detailed introduction to animal viruses may be found in Fenner et al. (1974).

14.1.1. Animal virus classification

The classification in Table 14.1 shows 16 groups of animal viruses and is based on (a) the nature of the nucleic acid, (b) the structural symmetry of the virus particle, (c) the presence of an envelope, and (d) the size of the virion. The classification is simplified from that of Fraenkel-Conrat (1974) and resembles that of Wildy (1971), but details of classifications differ as precise taxonomic relationships have not been established.

14.1.2. Precautions to be taken when using virus infected cells

Viruses produce cytopathic effects on cells and are the agents for many diseases in humans and other animals. In addition, many viruses (e.g. oncorna viruses, herpes type II, adenovirus and polyoma and SV40) are believed to be agents responsible for tumour formation in animals. Moreover, due to their ability to pass through bacteriological filters it is difficult to exclude viruses from uninfected cell cultures if the viruses are present in suspension in the air of the culture room. For these reasons it is recommended practice to take special precautions when using viruses.

The precautions suggested below are of a general nature only and should be used when no particular hazard is expected (i.e. for viruses not listed in the Godber report on dangerous pathogens, 1975). However, where particular hazards are known extra precautions may be required, and the investigator should refer to the above report which publishes a code of practice for use in laboratories holding category A pathogens, i.e. those so dangerous as to present great risks to the health either of laboratory workers or of the human or animal communities such that material containing live organisms should not be accepted knowingly or held at all in the United Kingdom without authorisation. Viruses which present parti-

Group	Nucleic acid		Average MW × 10^6	Virion diam. (nm)	Shape	Symmetry	± env.	eg.
Parvo	DNA	ss	2	20	Spherical	Icosahedral	–	MVM
Papova	DNA	ds	3–5	45–55	Spherical	Icosahedral	–	SV40
Adeno	DNA	ds	20–25	70–80	Spherical	Icosahedral	–	Ad 2
Herpes	DNA	ds	54–92	100–150	Roughly spherical	Icosahedral	+	HSV
Pox	DNA	ds	160–200	300 × 240 × 100	Brick-shaped	Complex	+	Vaccinia
Picorna	RNA	ss	~2	20–30	Spherical	Icosahedral	–	Polio
Toga or encephalo	RNA	ss	2–3	50–70	Spherical	Cubic	+	Sindbis
(Ortho)myxo	RNA	ss	~3	80–120	Roughly spherical	Helical	+	Flu
Corona	RNA	ss	~3	80–120	Roughly spherical	Helical	+	Murine hepatitis
Paramyxo	RNA	ss	7.5	100–300	Pleomorphic	Helical	+	NDV
Rhabdo	RNA	ss	6	175 × 75	Bullet shaped	Helical	+	VSV
Arena	RNA	ss		85–120	Roughly spherical	Helical	+	Lassa
Leuko (oncorna or retro)	RNA	ss	10–12	100–120	Roughly spherical	Complex	+	RSV
Reo	RNA	ds	10	70–90	Spherical	Icosahedral	–	Reo

Based on Fraenkel-Conrat (1974). ss and ds refer to whether the nucleic acid is single stranded or double stranded. ± env. indicates whether or not the virus is enveloped. The abbreviations for the viruses can be found in Appendix 6.

cular hazards include Newcastle disease virus; foot and mouth disease virus, vesicular stomatitis virus, smallpox virus, rabies virus, herpes type B virus, etc. However, it cannot be assumed that even viruses such as SV40 present no risk to the human population and the various hazards which may arise in biological laboratories are considered in a book published by Cold Spring Harbor Laboratories (Hellman et al., 1973).

- A special room, or suite of rooms, should be kept for the transfer and growth of cells infected with virus.
- No live virus should be removed from this area unless enclosed in a tightly closed container. Precautions should be taken to ensure that the outsides of all vessels are free of virus particles.
- Special protective clothing (lab. coats) should be available for use in the virus suite. After use these should be put in special bags for autoclaving (Sterilin Ltd.; Appendix 3).
- All medium and glassware that has been in contact with virus should be treated with chloros (§ 4.1.1) before removal from the virus suite.
- All plasticware should be sealed in special bags for autoclaving (Sterilin Ltd.; Appendix 3).
- Equipment for the storage and processing of virus material should be located within the virus suite.
- A double ended autoclave should be available so that laboratory workers are protected from hazardous materials.

14.2. Virus production

In order to study the virus growth curve a one-step growth cycle is performed. A high multiplicity of infection (m.o.i.) is used to ensure every cell is infected – usually 10 plaque forming units (p.f.u.) per cell is adequate. For virus production, however, the infection is prolonged under conditions where secondary infection can occur and a low m.o.i. is recommended, especially where there is a tendency for defective virus particles to be produced.

14.2.1. Procedure for production of herpes simplex, pseudorabies or EMC virus

- Seed 12 rotating Winchester bottles each with 2×10^7 BHKC13 cells in ETC_{10} (Eagle's medium containing 10% tryptose phosphate broth plus 10% calf serum) and grow at 37°C for 3 days.
- Infect each bottle with 1 plaque forming unit (p.f.u.) virus per 300 cells in 20 ml serum-free Eagle's medium.
- Incubate at 37°C for 1 h to allow the virus to absorb to the cell. Add 50 ml ETC_5 and incubate at 37°C for 2 days (or 32°C for 3 days) for herpes simplex virus, 27–36 h for pseudorabies virus or 24 h for EMC. If medium becomes acid add bicarbonate after 1 day.
- Harvest by shaking to dislodge the cell sheet into the medium. (If the cells do not come off the glass some versene may be added but NOT trypsin.)
- Centrifuge at 400 g for 10 min to pellet cells.
- Decant supernatant and spin at 20,000 g (100,000 g for EMC) for 90 min; discard this supernatant into chloros and resuspend the pellet in Eagle's MEM supplemented with 10% tryptose phosphate broth and 5% calf serum (Appendix 1) (2 ml/bottle). Dispense aliquots into bijoux bottles and store at −70°C (SUPERNATANT VIRUS).
- Disrupt the cell pellet by sonication or freeze–thawing 3 times in an ice/alcohol bath. Spin out the debris at 400 g for 10 min and dispense aliquots of this preparation at −70°C (CELL ASSO-CIATED VIRUS).
- Both batches of virus should be tested for bacterial contamination (Chapter 9) by inoculating an aliquot of virus into brain-heart infusion broth (leave at 37°C for 1 week) and into Saboraud fluid medium (32°C for 1 week) (Appendix 4). Virus may be purified from these preparations by sedimentation through solutions of high density, e.g. saturated KBr or RbCl (Black et al., 1964) when bands corresponding to virions and empty particles are evident.

14.2.2. Procedure for production of SV40 virus

a. Seed 20–30 9-cm dishes with 2×10^6 BSCl cells in EFC_{10} (Eagle's medium containing 10% foetal bovine serum).

b. After one or two days growth infect the cells with 0.01 p.f.u./cell (preferably use virus picked from plaques (see § 14.3.1) as this is known to be non-defective).

c. After growth for a week to 10 days (i.e. when most of the cells show a strong CPE) scrape the cells into the growth medium. Stand at 4°C for 1 h and then centrifuge at 15,000 g for 30 min.

d. Resuspend the cells in 20 ml PBS-A and add pancreatic DNase I (40 μg/ml) and pancreatic RNase A (12 μg/ml).

e. Freeze–thaw 3 times or sonicate and then incubate at 37°C for 30 min.

f. Add deoxycholate to 0.15% and stand at room temperature for 90 min. Centrifuge at 15,000 g for 30 min at 20°C.

g. Remove the supernatant into a Spinco SW27 tube and underlay with 10 ml saturated KBr in PBS-A. Centrifuge 23,000 r.p.m. for 3 h at 20°C. Two bands should be visible in the KBr solution: a lower band containing virions and an upper band of empty particles.

h. Collect the virion band by puncturing the tube and dialyse against PBS-A (2×1 h).

i. Add solid CsCl to a density of 1.34 g/ml and centrifuge 3 ml aliquots (overlayered with paraffin oil) in the Spinco SW50.1 rotor for 18 h at 40,000 r.p.m. at 20°C. The virions band with a buoyant density of 1.34 g/ml.

j. Collect fractions, measuring the buoyant density from refractive index, and the absorbance at 260 and 280 nm to detect the virions and empty particles (which will have relatively low absorbance at 260 nm).

k. Pool the virion containing fractions and dialyse against PBS at 4°C for 2 h.

l. Sterilise by filtration through a 0.22 μm filter if required for further infections.

Although this procedure is necessary for production of samples of pure virions it is usually unnecessary if a viral preparation is only required for subsequent infections. In such circumstances the cells should be harvested aseptically and processed to step e, omitting the DNase and RNase. The debris from the disrupted cells is pelleted at 15,000 g for 30 min and the supernatant used as a source of virus. It should be tested for bacterial contamination with brain-heart infusion broth and Saboraud fluid medium (Appendix 4). The virus should be stored at $-70°C$ at about 10^{10} p.f.u./ml.

14.2.3. One-step growth curve of SV40

This may be studied in 9-cm dishes or the wells in a tissue culture tray.

- Seed with 2×10^6 (9-cm dish) or 10^5 (TC well) cells in 10 ml or 0.5 ml Eagle's medium containing non-essential amino acids, penicillin–streptomycin and 10% foetal bovine serum (Appendix 1).
- After a day or two of growth remove the medium and add SV40 to give 1 p.f.u./cell (2 ml/dish; 0.1 ml/well).

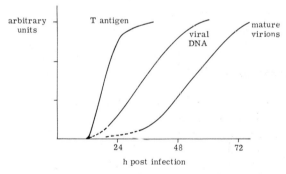

Fig. 14.1. Time course of SV40 infection. Monkey cells in monolayer culture, infected with SV40 at 1–10 p.f.u. per cell. T antigen may be detected by immunofluorescence, viral DNA synthesis by labelling with [^{3}H]thymidine followed by separation of viral DNA (by Hirt extraction, SDS sucrose gradient centrifugation or agarose gel electrophoresis) and mature virions by infectivity using a plaque assay. (Data from Tooze, 1973; Girard et al., 1975; and Basilico and Zouzias, 1976.)

- Incubate at 37°C for 60–90 min to allow time for virus adsorption, overlay with Eagle's medium containing 5% foetal bovine serum (8 ml/dish; 0.4 ml/well).

Viral DNA synthesis is maximal on the 2nd day post infection. It may be detected on the first day and it is associated with – or slightly preceded by – the production of T-antigen. Mature virions are detected after a short lag (Fig. 14.1).

14.2.4. Sendai virus – production and inactivation

14.2.4.1. Production
Sendai virus is grown in the allantoic cavity of hens' eggs. The method is described by Harris and Watkins (1965).

- Take infected allantoic fluid at 8000 haemagglutination units/ml (§ 14.3.3) and dilute 1 in 10^4 with PBS.
- Inject 0.1 ml into the allantoic cavity of 10–11-day-old fertile hens' eggs.
- Incubate at 37°C for 3 days and then overnight at 4°C.
- Collect the allantoic fluid and centrifuge at 400 g for 10 min.
- Determine the haemagglutination titer (§ 14.3.3).
- Centrifuge 30,000 g for 30 min and resuspend the pellet in 1/10 the original volume of Hanks' BSS.
- Store at −70°C.

14.2.4.2. Inactivation by UV

- Place 1 ml concentrated virus on a watch glass and expose for 3 min to UV light from a 15 W germicidal lamp at incident radiation of 3000 erg.cm^{-1} s^{-1}. Mix after 1 min and 2 min (see § 12.4.2).

14.2.4.3. Inactivation by β-propiolactone (BPL) (Neff and Enders, 1968)

- Prepare a 10% aqueous solution of BPL immediately before use.
- Dilute to 1.3% with isotonic saline containing 1.68% NaHCO$_3$.
- Add 1 part diluted BPL to 9 parts Sendai virus (haemagglutination titer between 1 : 2000 and 1 : 10,000 (§ 14.3.3).

- Shake in a tightly sealed container at 4°C for 10 min and then keep at 37°C for 2 h shaking every 10 min.
- Keep overnight at 4°C to ensure complete hydrolysis of the BPL.
- Inactivated virus may be stored in the presence of 0.5% serum albumin at −65°C for 5 weeks.

14.3. Virus detection

The presence of virus may be recognised and in some cases quantitated by a number of tests, e.g.

a) production of characteristic effects on cells which may be used in a plaque assay (see below),

b) production by infected cells of virus particles which may be identified using the classification in Table 14.1, i.e. nature of nucleic acid, symmetry of particle,

c) production by infected cells of viral nucleic acid which may exhibit a characteristic density on buoyant analysis or a characteristic size on agarose gels or sucrose gradients (Kaplan, 1969; Tegtmeyer, 1972),

d) production by infected cells or transformed cells of characteristic antigens which may be recognised by staining with fluorescent antibody (§ 14.3.2) or which cause changes in the cell membrane leading to haem adsorption (Deibel and Hotchin, 1961),

e) the presence of antigens on virus particles may lead to haemagglutination reactions which can be readily quantised (§ 14.3.3),

f) production by infected cells of characteristic enzymes or enzymes with properties which readily distinguish them from the corresponding host cell enzymes (Keir et al., 1966; Baltimore and Smoler, 1971; Harada et al., 1975).

Some of these tests will now be described. As the number of possible viruses and tests is endless, those chosen simply reflect those with which I am familiar.

14.3.1. Plaque assay

The method depends on infecting a small number of cells in a confluent monolayer. The virus produced in the infected cells will move laterally to infect adjacent cells and various techniques are used to prevent further spreading. The degenerative effect on the cells spreads until a visible area of dead cells (a plaque) is apparent. Staining of the cell sheet makes the colourless plaque more easy to see.

Usually cell cultures are infected with different dilutions of the viral preparation covering a range of 10^4. Thus a viral stock will be diluted in tenfold increments.

14.3.1.1. Viral dilution

Set up a series of tubes containing 0.9 ml BSS or PBS or Eagle's medium without serum and to the first tube add 0.1 ml virus stock. Mix the contents and remove 0.1 ml to the second tube and so on. A solution is required which contains 200–400 infectious units per ml and usually the 10^{-5} to 10^{-9} dilutions are assayed.

14.3.1.2. Suspension assay

Herpes and pseudorabies viruses will infect BHK C13 cells in suspension.

a. Prepare a cell suspension containing 3×10^6 cells/ml.

b. Add 4 ml of suspension to each of 5 universals.

c. Add 0.8 ml of the virus dilutions to each universal.

d. Shake gently on the mechanical shaker at 37°C for 20 min.

e. After 20 min add 12 ml Eagle's + 10% tryptose phosphate broth + 10% calf serum (for pseudorabies virus) or Eagle's + 10% tryptose phosphate broth + 5% human serum (for herpes virus) to each universal.

f. Mix and plate 4 ml/50 mm Petri dish.

g. Incubate in a CO_2 incubator at 37°C. After 2 h add 25 μl of a heparin solution (10 mg/ml) to the pseudorabies infected cultures.

h. After 28 h (pseudorabies) or 48 h (herpes) remove the medium from the dishes (put it straight into chloros).

i. Fix the infected monolayer with 10% formol saline for 10–20 min.
j. Remove formol saline and stain with Giemsa for 10–20 min.
k. Wash off stain gently under tap water and count the plaques using a low power microscope.

14.3.1.3. Monolayer assay

For herpes and pseudorabies virus this assay is similar to the suspension assay except that infection occurs on a confluent monolayer (Fig. 14.2).

Fig. 14.2. Plaques of herpes simplex virus on a BHK/C13 cell layer. A confluent layer of BHK21/C13 cells was infected with a herpes simplex virus type 2 preparation at different dilutions. On the left few plaques are visible but on the right there are many plaques. (With thanks to Dr. J.B. Clements.)

– Set up 20 dishes (50 mm) with 3×10^6 BHK C13 cells per dish. Grow at 37°C for 24 h.
– Remove the medium from the dishes and inoculate 0.2 ml of each virus dilution onto each of 4 dishes (include a control mock infected with 0.2 ml BSS).
– Allow the virus to adsorb at 37°C for 1 h in the CO_2 incubator. Rock the dishes from time to time to spread the virus over the surface.

- After 1 h add 5 ml Eagle's medium + 10% tryptose phosphate broth + 10% calf serum (pseudorabies) or 5% human serum (herpes) and return to the incubator.
- After 2 h add heparin to the pseudorabies virus infected cultures.
- Continue as for suspension assay, stages h–k.

14.3.1.4. Agar overlay assay

- Set up 20 × 50 mm dishes with 2–4 × 10^6 BHK21/C13 (EMC virus) or BSCl (SV40) cells per dish and grow for 24 h at 37°C to obtain confluent monolayers.
- Remove the medium from each dish and inoculate 0.2 ml of each virus dilution onto each of 4 dishes (also inoculate controls with BSS).
- Allow the virus to adsorb at 37°C for 1 h in the CO_2 incubator. Rock the dishes from time to time to spread the virus over the monolayer.
- After adsorption add 5 ml agar overlay medium to each dish. This is made up as follows:

  Mix Eagle's MEM*   75 ml
       Noble agar 2.5%   25 ml (46°C; see below)
       Calf serum (for EMC virus)
       or
       Foetal calf serum (for SV40)   5 ml
       DEAE dextran (0.5%)   1 ml

  and equilibrate at 46°C. It is important to maintain the temperature between 45°C and 50°C. The agar sets shortly after pouring onto the plates at 37°C.

  *The Eagle's MEM should be made up at 1.3 times the normal concentration (Eagle's × 1.3).

- Incubate at 37°C for 24 h (EMC virus) or 6 days (SV40).
- Add 2 ml neutral red overlay medium (see below) directly onto the agar and allow to solidify in total darkness.
- Return to the incubator for 2–3 h (or even for several days if the plaques are too small for counting).

– Plaques may be seen developing and can be counted using a plate microscope or colony counter.

To make up 2.5% Noble agar (Difco Labs.) add 25 g agar powder to 500 ml distilled water and make up to 1 l. Place the flask in a container of boiling water until the agar dissolves. Dispense 25 ml amounts into universal containers with metal caps and autoclave at 15 lb pressure for 15 min. Store at room temperature.

To make the neutral red overlay medium first dissolve neutral red to 0.4% in distilled water (heat if necessary) and filter through Green's Filter Paper No. 904½ (Appendix 3). Bottle in 20 ml amounts and sterilise by autoclaving at 15 lb pressure for 15 min. Store at room temperature. For use in the overlay medium add 2.5 ml to 75 ml Eagle's × 1.3 and 25 ml 2.5% agar.

14.3.2. Fluorescent antibody techniques

In the direct method an antibody raised against a viral antigen is coupled with a fluorochrome and used to stain infected cells. A positive reaction (a yellow-green colour in the fluorescence microscope) indicates the presence in cells of viral antigens. The method may therefore be used to detect cell transformation when the antibody is directed for instance against the SV40 early antigen or the production of mature virus if the antibody is directed against capsid protein.

In the indirect method the antibody is not coupled with the fluorochrome. Rather, after antigen and antibody have reacted in the fixed cell preparation, an anti-gamma globulin antibody conjugated to the fluorochrome is added. This method is more sensitive and eliminates the necessity for conjugating each precious antibody with fluorescent dye. Thus one fluorescent anti-rabbit antibody (raised in sheep against rabbit gamma globulins) will pick out and label any antibody raised in rabbits which has reacted with a viral antigen in the infected or transformed cells.

An even more indirect but reputedly far more sensitive test which does not require the use of a fluorescence microscope is the peroxidase anti-peroxidase (PAP) method of Sternberger et al. (1970). Here, after the antigen has been reacted with a rabbit antibody preparation this is conjugated to sheep anti-rabbit IgG which in turn is conjugated to a complex of horseradish peroxidase and rabbit anti-horseradish peroxidase. This then reacts with diaminobenzidine and hydrogen peroxide when a brown colour indicates the presence of the antigen.

The merits of the various systems have been discussed by Taylor (1978) who gives details of various immunoperoxidase techniques.

14.3.2.1. Preparation of antisera

Cells or cell extracts when injected into animals cause the production of antibodies. The procedures are detailed by Clausen (1969) in this series of laboratory manuals, and the animals usually used are rabbits, mice, hamsters, etc. Antisera to viral antigens are best raised in a syngeneic host (i.e. a host of the same genotype as that of the cells in which the virus is grown in vitro) if this is available, e.g. injection of SV28 cells (SV40 transformed BHK21/C13 cells) (Wiblin and Macpherson, 1972) into Syrian hamsters leads to tumour production and antibody to T-antigen is present in the serum.

14.3.2.2. Preparation of globulin fraction

A crude fractionation of antisera is usually carried out to remove most of the albumin. Either add slowly at 4°C an equal volume of cold saturated ammonium sulphate solution (adjusted to pH 7 with ammonia) and stir for a further 30 min or dialyse overnight at 4°C against 50 volumes of 50% saturated ammonium sulphate.

Centrifuge at 1200 g for 20 min at 4°C and dissolve the precipitate in PBS-A and dialyse against PBS-A (3 changes of 50 volumes) to remove the ammonium sulphate. For full details of purification and characterisation, see Clausen (1969; this series).

Antisera to gamma globulins of various animals may be prepared similarly but are available from commercial suppliers, e.g. Flow Laboratories (Appendix 3).

14.3.2.3. Conjugation of antisera with fluoroscein or rhodamine
These are the two common fluorochromes and they are covalently
linked to proteins by using activated forms (e.g. the isothiocyanates).
Fluoroscein isothiocyanate (FITC) and tetramethylrhodamine isothiocyanate (TRITC) are commercially available from, for example,
Sigma Chemical Co. (Appendix 3).
  More uniform reaction is obtained if the isothiocyanates are first
absorbed onto Celite. FITC–Celite is available from Sigma Chemical
Co. Ltd., but can be easily prepared:
  a) dissolve 25 mg FITC in 25 ml acetone
  b) add 500 mg Celite and stir for a few minutes at room temperature
  c) remove the acetone under vacuum and store dry in a desiccator.

– Mix 0.5 ml globulin fraction with an equal volume of carbonate–
  bicarbonate buffer, pH 9.3 (4.4 ml 5.3% sodium carbonate plus
  100 ml 4.2% sodium bicarbonate). Add 15 mg FITC–Celite and
  seal to prevent loss of CO_2. Mix gently for 30 min at room
  temperature.
– Remove the Celite by centrifugation (800 g, 10 min) and separate
  the conjugated protein from excess dye on a Sephadex-G25 column
  (1 × 10 cm) eluted with PBS-A.

Antisera (and conjugated antisera) may be stored at −20°C or
kept at 4°C in the presence of merthiolate (1 part in 10,000) which
acts as a preservative.

14.3.2.4. Staining techniques

Direct method
a. Grow cells on coverslips and infect with virus (say SV40).
b. Remove the coverslips from infected and mock infected cultures
   and, holding them in forceps or in a micro coverslip carrier, dip
   them into two washes of PBS-A at room temperature and dry
   them by dipping into two washes of acetone at 0°C. Air-dry
   (cells up) on filter paper (take care to identify individual coverslips
   by their position).

c. Place coverslips in a 5 cm dish containing wet filter paper. It is often convenient to stand them on small rubber stoppers.
d. Flood each coverslip with 2 drops FITC conjugated SV40 T antigen hamster antiserum (§ 14.3.2); place a lid on the dish and leave at room temperature for 45 min.
e. Wash coverslips in three changes of PBS-A and mount by inverting onto a slide using glycerol as mounting medium.

Indirect method
Repeat steps a–d of the direct method above.
d′. Flood each coverslip with 2 drops unconjugated SV40 T antigen hamster antiserum (§ 14.3.2); place a lid on the dish and leave at room temperature for 30–45 min.
e′. Wash coverslips in three changes of PBS-A and drain for 10 min at room temperature; replace in dish.
f. Flood with 2 drops FITC conjugated goat anti-hamster antiserum; replace the lid and leave at room temperature for 30 min.
g. Wash coverslips in three changes of PBS-A and mount by inverting onto a slide using glycerol as mounting medium. Examine the coverslips using a fluorescence microscope.

14.3.3. Haemagglutination

Many viruses have antigens which adsorb to and cross-link red blood cells. When many virus particles and red cells interact a network is formed and the suspension agglutinates, i.e. the red cells are precipitated. There is some specificity as to which viruses will agglutinate red cells from which animal, but once the combination is known, haemagglutination forms a quick assay for virus titre.

Blood is collected in a heparinised syringe and the red cells washed three times in 0.85% saline by sedimentation at 200 g for 10 min and resuspension. They are finally resuspended in 200 volumes 0.85% saline.

The assay is most conveniently carried out in a microtitre plate (Fig. 3.2) and the dispensing may be done by hand or using automatic equipment available from Titertek (Flow Labs. Ltd., Appendix 3).

25 μl of 0.85% saline or **PBS** is dispensed into a series of wells in the microtitre plate and 25 μl of virus suspension is added to the first well and mixed (1 in 2 dilution). 25 μl is then removed from the first well and mixed in the second well (1 in 4 dilution) and so on to give dilutions up to 1 in 2048. 25 μl of red blood cell suspension is now added, mixed, and left, either at 4°C, room temperature or 37°C until haemagglutination has occurred (1–2 h).

In wells where positive agglutination has occurred the red cells have an irregular outline while in negatives and in the control the cells settle in a compact dot in the bottom of the well. The dilution in the last well showing a positive reaction is considered the titre and this dilution has 1 haemagglutinating (HA) unit. The next dilution has 2 HA units and so on.

14.4. Viral transformation of cells

When a virus penetrates a cell instead of causing a productive infection it may transform that cell. For this to happen its genome (or a part of its genome) must integrate into the cellular chromosomes. The site of integration does not appear to be unique and often several copies of the viral DNA are integrated.

The following characteristics have been noted in cells transformed by SV40 or polyoma virus (Tooze, 1973).

Growth: high or indefinite saturation density
reduced serum requirement
growth in agar or Methocel (see Appendix 6) suspension
tumour formation on injection into susceptible animal
not susceptible to contact inhibition of movement
growth in disoriented manner
growth on monolayers of normal cells

Surface: increased agglutinability by plant lectins
changes in glycoprotein and glycolipid composition
foetal antigens revealed
tight junctions missing
transplantation antigen present

Evidence of virus: virus specific antigens present
                    viral DNA sequences detected
                    viral mRNA present
                    virus can sometimes be rescued

14.4.1. Methods of transformation

There are at least four methods of transforming cells with viruses.

A. The simplest is that of Macpherson and Montagnier (1964). The cells are infected with the virus in suspension and then plated in 0.33% agar or 1.2% Methocel (Appendix 6) when transformed cells form large colonies and untransformed cells do not grow.

Prepare a basal layer of 7 ml of medium (Eagle's Glasgow modification containing tryptose and calf serum) containing 0.5% Difco Bacto agar in a 6 cm dish.

Infect a cell suspension for 1 h at 37°C and add $10^3 - 5 \times 10^5$ cells in 1.5 ml medium containing 0.3% agar or 1.2% carboxymethyl cellulose (Stoker, 1968) to the preset basal layer.

Incubate 7–10 days at 37°C in a CO_2 incubator when transformed cells alone form colonies.

B. Expose subconfluent monolayers of cells to the virus and grow for 2–3 weeks when dense clones of transformed cells grow out of the untransformed monolayer.

C. Infect cells and plate out at a low density in low serum medium. Many cells may divide once or twice (abortive infection) but the stably transformed cells will form colonies in the absence of serum factors (Smith et al., 1970).

D. The above methods rely on the selection of the transformed cells making use of one of the properties listed above. Stoker and Macpherson (1961) exposed confluent monolayers or cell suspension to virus and then plated the cells at low density. They then picked out transformed colonies by appearance in the absence of any selective pressure.

CHAPTER 15

Differentiation in cell cultures

The study of differentiation in multicellular animals is beset by problems arising from the complex and poorly understood interactions of the various tissues. These interactions can be limited or eliminated in cell culture and a number of systems are described below which are leading to an understanding of the mechanisms of cell and tissue differentiation.

15.1. Erythroid differentiation of Friend cells

The process of erythropoiesis has been reviewed by Harrison (1976, 1977) and Orkin (1978). Erythroid cells together with the other blood cells are derived from a common haematopoietic stem cell. After commitment to the erythroid lineage the stem cells proliferate for a few generations when they become sensitive to the hormone erythropoietin which increases the proliferation of committed erythroid stem cells and proerythroblasts which then differentiate into mature erythroid cells containing haemoglobin.

Studies with normal erythropoietic systems are hampered by the difficulties of obtaining sufficient erythroid cells of specific developmental stages.

Friend cells are murine, virus-induced, erythroleukaemic cells which grow in culture in suspension and exhibit a limited degree of differentiation along the erythroid line (Friend et al., 1966; Patuleia and Friend, 1967). The target cell for the Friend virus is probably a late committed erythroid stem cell. Under normal growth conditions only 1–2% of the cells stain with benzidine (haemoglobin-positive).

They appear to be arrested at the proerythroblast stage and are not responsive to erythropoietin. However, on treatment with dimethyl-sulphoxide (2%) after a 3-day lag the proportion of benzidine-positive cells rises to 96% on the 5th day (Friend et al., 1971), i.e. they undergo a series of changes characteristic of normal erythroid cell differentiation.

Marks and Rifkind (1978) have reviewed the characteristics of the induced differentiation and listed the large variety of inducers (e.g. butyric acid, hemin, ouabain) which may be used in place of dimethylsulphoxide. However, not all these inducers produce the same effects in all the clones tested and some DMSO resistant variants exhibit different phenotypes (Harrison et al., 1978).

The Friend cells have been used, therefore, as an ideal in vitro system in which to study differentiation and it has been shown that the action of DMSO leads to cessation of cell division and production of globin messenger RNAs (Harrison, 1976; Minty et al., 1978). This induction of globin mRNA synthesis appears to be blocked specifically by hydrocortisone (Scher et al., 1978) and is dependent on some event which occurs in the G1 phase in synchronised cells (Gampari et al., 1978) though whether or not DNA synthesis is essential appears open to question (Harrison, 1976).

In the absence of inducer, Friend cells continue to divide normally but in the presence of DMSO although a tenfold increase in cell number may occur cell viability is markedly reduced.

15.1.1. Induction of globin synthesis in Friend cells

Place 10^6 cells in a 6-cm diameter Petri dish containing 10 ml Eagle's BME supplemented with 15% foetal calf serum and add reagent grade DMSO (unsterilised or sterilised by autoclaving in small volumes) to 2%. After 5–6 days in DMSO the cells may be washed in PBS and then lysed in distilled water. By reading the optical density of the clarified supernatant at 415 nm the amount of haemoglobin present may be estimated. Alternately the cells may be grown in the presence of ^{59}Fe and radioactive haem extracted and counted to give a measure of haemoglobin synthesis (Freshney, 1975).

15.2. Skin and keratinocytes

The differentiation of the epidermal cells into a basal layer of dividing cells and an upper layer or layers of cells which become keratinised and eventually sloughed off appears a relatively simple system and it is made even simpler by the ability to grow keratinocytes in culture.

Such an in vitro system can be used not only to study the differentiation of skin and diseases which affect the skin but may also be used to test drugs which affect human epidermis and possibly to produce large amounts of epidermal cells which could be used in skin grafting.

Epidermal cells depend for their maintenance and growth on the presence of fibroblasts or their products and when the two cells are present in optimal proportions growth and differentiation proceed very well in culture. However, their lifetime is restricted to 20–50 generations (Hayflick and Moorhead, 1961), which is very much shorter than the in vivo lifetime of a cell of the basal layer of the human epidermis which undergoes 30 divisions per year. The culture lifetime can be increased to about 150 generations by addition of epidermal growth factor (Rheinwald and Green 1977; see § 7.7.2) and cAMP levels may also play an important role in growth regulation (Green, 1978).

Described below is a method for growing keratinocytes from skin biopsies using a feeder layer (Rheinwald and Green, 1975). The feeder cells may be 3T3 or BHK21 cells treated with gamma rays or with mitomycin C as described in § 8.1.5.

– Dissect the skin biopsies (see § 6.3) into Dulbecco's MEM with 10% calf serum (Appendix 1) and mince finely.
– Stir, in about 10 ml 0.25% trypsin in PBS-A at 37°C.
– Every 30 min allow lumps of tissue to settle for 1 min and withdraw the supernatant, replacing it with fresh trypsin.
– Sediment the cells from the supernatant and resuspend in growth medium containing 20% foetal calf serum and hydrocortisone

(0.4 μg/ml) (this accelerates growth and makes colony morphology more orderly).

- Mix with a suspension of feeder cells and distribute the cells into dishes. The proportions of the two cell types affect the character of the subsequent growth.
- The dishes should be left undisturbed as epidermal cells take 2–3 days to attach firmly. Thereafter the medium should be replaced twice weekly. The growth of fibroblasts is largely suppressed but the epidermal cells form colonies of keratinocytes which push back the feeder layer at the periphery.
- To subculture the keratinocytes first treat the culture with 0.02% EDTA for 15 s and pipette vigorously to selectively remove the fibroblasts.
- Then disaggregate the keratinocyte colonies with 0.02% EDTA: 0.05% trypsin (1 : 1). The keratinocytes may be replated with a fresh feeder layer if required.

If the cultures are not subcultured when they become confluent then cells are shed from an upper layer while the basal layer continues to divide (Green, 1977). Thus the situation closely resembles the stratum corneum in vivo where the basal cells multiply and the non-basal cells undergo differentiation.

15.3. Teratocarcinoma cells

Teratomas and teratocarcinomas have been recently reviewed by Hogan (1977) and Illmensee and Stevens (1979). Teratomas are tumours arising in the ovary or testis or in early embryos transplanted into extrauterine sites. They consist of a mixture of tissues among which are recognisable skin, nerve, muscle, cartilage, etc. If, interspersed among the recognised tissues, there are undifferentiated embryonal cells, the tumour is called a teratocarcinoma and is recognisable by a fast growth rate and by its ability to be transplanted into syngeneic animals (i.e. animals of the same genotype). Incorporation of mouse teratocarcinoma cells into the cell mass of early mammalian embryos can give rise to completely normal chimeric adults in which the teratocarcinoma cells have been incorporated into many different tissues (Mintz and Illmensee, 1975; Papioannou et al., 1975). Thus in vivo these cells are able to participate in normal morphogenesis and differentiation, and a considerable amount of work has recently gone into mimicking these events in vitro. Thus a number of cell lines have been established which can be triggered to differentiate rapidly and in a predictable manner following stimulation.

Two methods have been developed to establish mouse teratocarcinoma cells in culture.

1. Tumour cells are allowed to attach to gelatin-coated tissue culture dishes (see § 2.3.2) when many different kinds of cells including embryonal cells grow out. These latter cells may eventually outgrow the differentiated cells and they may then be cloned (Chapter 8) (Rosenthal et al., 1970; Bernstine et al., 1973). Lines isolated in this way tend after some time to lose their ability to differentiate.

2. Tumour cells (ascites or dissociated solid tumours) are plated out on a monolayer of feeder fibroblasts (see § 8.1.5). After several days many different sorts of differentiated cells may be seen and also nests of embryonal cells. On repeated subculture these cells will predominate, but it is perhaps easier to isolate them manually and reseed on a feeder layer (Martin and Evans, 1975a).

Once isolated the undifferentiated cells remain homogeneous as long as they are subcultured using 0.25% trypsin, 0.05 mM EDTA in PBS (Chapter 5) into Dulbecco's modified Eagle's medium (Appendix 1) supplemented with 5–20% bovine serum before they become confluent.

If the teratocarcinoma cells become confluent they will begin to differentiate into many different cell types, but this process is not so dramatic as the differentiation which occurs if cells isolated by method 2 are transferred to a vessel without a feeder layer. Then the cells attach very poorly and form clumps in suspension. These clumps remain healthy and quickly differentiate (Evans, 1972) to form an outer layer of endoderm cells. The presence of endoderm can be shown by assaying for the serine protease plasminogen activator which is a marker typical of endoderm cells (Strickland et al., 1976). These aggregates are known as embryoid bodies and after 2–3 days in culture they develop a fluid filled cyst on one side and other signs of differentiation may be seen. If the aggregates are allowed to reattach to tissue culture dishes several kinds of differentiated cells grow out including beating muscle, nerve and glandular tissue (Martin and Evans, 1975b).

Some teratoma cell lines isolated by method 1, e.g. the F9 cell line, have almost lost their capacity to differentiate. However, Strickland and Mahdavi (1978) have shown that differentiation may be induced with retinoic acid at 10^{-9} M. Again endoderm formation is detected by formation of plasminogen activator.

15.4. Myeloma culture and antibody production

In vivo mature differentiated lymphocytes normally synthesise and secrete immunoglobulin (Ig) molecules. Myelomas (tumours) of plasma cells (plasmacytomas) generally synthesise and may secrete one species of immunoglobulin (a myeloma protein) though myelomas derived from early stages of lymphocyte differentiation may synthesise more than one Ig species (e.g. the line SAMM 368). As myelomas arise from previously stimulated lymphocytes the specificity of the antibody (the Ig) is unknown. A frequently used myeloma line, MOPC 315, has been shown to produce an immunoglobulin with affinity for several of the large number of antigens screened. The nature and usefulness of myelomas have been reviewed by Potter (1972, 1975, 1976) and Rabbitts and Milstein (1977).

Plasmacytomas may arise spontaneously or be induced, usually in inbred strains of mice (BALB/c or NZB), by injection of mineral oil or plastic discs. The tumours are normally passaged as solid or ascitic tumours in mice but may be adapted to in vitro culture, where they are generally grown in suspension although under these conditions they tend to lose some of their Ig synthetic activity.

The cells grow rapidly and are hyperdiploid. They are grown in RPM1 1640 (Chapter 7 and Appendix 1) supplemented with 10% foetal calf serum and 1% glutamine. They adhere loosely to the surface but may be grown in a stationary culture or in roller bottles at $2–8 \times 10^5$ cells/ml. They may be dislodged from the substratum by shaking the bottle.

Antibodies are powerful tools when it is necessary to characterise the surface proteins (antigens) of a cell and can be used to mark subpopulations of cells; important particularly with regard to specific cell surface phenomena.

The main problem is to raise antibodies specific to individual cell surface molecules. If antisera are raised by immunising one species with cells or subfractions of cells from another a complex response is induced and the antisera require considerable fractionation and purification.

In an attempt to overcome this problem Kohler and Milstein (1975, 1976) have reported the results of fusing (Chapter 13) a mouse myeloma cell line with spleen cells taken from a mouse immunised with sheep erythrocytes. The hybrids were cloned and clones secreting anti-sheep erythrocyte antibodies selected. Williams et al. (1977) have fused mouse myeloma cells with mouse spleen cells immunised with rat thymocyte membranes and isolated 5 clones each producing a different antirat thymocyte antibody.

After fusion of 10^7 MOPC 21 derived myeloma cells (Horibata and Harris, 1970) with 10^8 spleen cells using polyethyleneglycol (Chapter 13) the culture was divided into the 48 wells of a tissue culture tray (Linbro BCL-5041 tray; see Appendix 3) and maintained in Dulbecco's MEM (Appendix 1) supplemented with 10% horse serum and HAT (§ 13.5) and maintained in an atmosphere of 10% CO_2 in air. The medium was tested periodically using a radioactive binding assay for the presence of anti-rat thymocyte antibodies and positive cultures were cloned using the soft agar technique (§ 8.1.4).

15.5. Differentiation of muscle cells

The transition from dividing myoblasts to multinucleate muscle fibres is one of the most striking examples of terminal differentiation which can occur in vitro. The biochemical changes result in the development of an excitable membrane, assembly of the contractile apparatus and the appearance of appropriate enzymes, and have been summarised by Buckingham (1977).

The added attraction of the transition from myoblast to myotube is the synchrony with which differentiation occurs in vitro. Myogenesis will occur in primary cultures of skeletal muscle (e.g. § 6.6) but can also be induced in diploid myoblast lines (Richter and Yaffé, 1970) which has allowed the selection of mutants (Chapter 13) that exhibit drug resistance or temperature-sensitive differentiation (Loomis et al., 1973; Somers et al., 1975). Holtzer et al. (1975) and Fiszman and Fuchs (1975) have developed a myoblast line transformed with a temperature-sensitive virus (§ 13.3 and § 14.4). At the permissive temperature the cells divide continuously but when the temperature is raised differentiation is induced.

Myoblasts may differentiate and fuse without undergoing DNA synthesis (Nadel-Ginard, 1978).

15.6. Differentiation of adipose cells

From the 3T3 cell line originally isolated from mouse embryos (Todaro and Green, 1963) Green and Meuth (1974) have isolated two clones which, on entering the resting state, accumulate large amounts of triglycerides in multiple cytoplasmic droplets. The resting state ensued as the cells reached confluence but could be induced synchronously by trypsinising the cells and inoculating a suspension into Eagle's medium supplemented with 20–30% calf serum and containing carboxymethyl cellulose (see § 14.4.2).

Although all the cells retained the ability to synthesise collagen (i.e. they are fibroblasts) only a proportion of the cells in a series of different subclones actually produce lipid and this fraction never reaches 100% even after several weeks of incubation of resting cells.

The subsequent ability of cells to induce triglyceride synthesis and storage is interfered with by incubation of growing (but not resting) cells with bromodeoxyuridine at 5 μM. This compound is known to interfere with several differentiating systems (Rutter et al., 1973) at concentrations which do not affect the rate of cell growth.

CHAPTER 16

Appendices

Media formulations

TABLE 1

Earle's and Hanks' balanced salt solutions (BSS) 10 × concentrated, without bicarbonate and glucose

	To make 10 litres			
	Earle's		Hanks'	
NaCl	680	g	800	g
KCl	40	g	40	g
MgSO$_4$.7H$_2$O	20	g	20	g
Na$_2$HPO$_4$.7H$_2$O	–		9	g
KH$_2$PO$_4$	–		6	g
NaH$_2$PO$_4$.2H$_2$O	14	g	–	
CaCl$_2$.6H$_2$O*	39.3	g	27.6	g
Phenol red (1%) – Table 7	150	ml	150	ml
Distilled water to	10	l	10	l

In some cases in Hanks' BSS half the MgSO$_4$.7H$_2$O is replaced by MgCl$_2$.6H$_2$O.
* This is best dissolved separately and added last with stirring.
Add 10 ml chloroform and store at 4°C in polythene containers.
CMF (calcium and magnesium free BSS) is Earle's BSS made up without the calcium and magnesium.

BSS may be required as a saline solution, for example for washing tissue minces in which case 50 ml of the 10 × concentrate is diluted with 450 ml distilled water.

246

However, if the BSS is to be used as a basis for a cell culture medium (when further ingredients will be added later) 50 ml of the 10 × concentrate is diluted with 350 ml or 400 ml distilled water. The dilution should be in 500 ml bottles which should be autoclaved at 15 lb pressure for 20 min. BSS may be stored at room temperature.

Before use add bicarbonate (20 ml 5.6% for Earle's BSS and 3 ml 5.6% for Hanks' BSS) or Hepes buffer (5 ml 1 M) to adjust the pH to 7.4; and glucose or other ingredients as desired.

TABLE 2

Dulbecco's phosphate buffered saline (PBS)

PBS solution A	1 litre	10 litres
NaCl	10 g	100 g
KCl	0.25 g	2.5 g
Na_2HPO_4	1.44 g	14.4 g
KH_2PO_4	0.25 g	2.5 g
pH 7.2		

Bottle in 160 ml or 400 ml amounts.
Autoclave at 15 lb pressure for 20 min to sterilise.
Store at room temperature.

PBS solution B	1 litre
$CaCl_2.2H_2O$	1.0 g

Bottle in 20 ml or 50 ml amounts.
Autoclave at 15 lb pressure for 15 min to sterilise.
Store at room temperature.

PBS solution C	1 litre
$MgCl_2.6H_2O$	1.0 g

Bottle in 20 ml or 50 ml amounts.
Autoclave at 15 lb pressure for 15 min to sterilise.
Store at room temperature.

To constitute PBS proper add 20 ml each of PBS-B and PBS-C to 160 ml PBS-A. Alternatively add 50 ml each of PBS-B and PBS-C to 400 ml PBS-A.

TABLE 3

Sodium bicarbonate (5.6%)

- Dissolve 56 g $NaHCO_3$ in distilled water. Add 1.5 ml 1% phenol red and make up to 1 l.
- Sterilise by filtration using a 0.22 μm membrane filter.
- Bottle in 20, 100, and 200 ml amounts.
- Cap the bottles very tightly using bottles with rubber lined metal caps if possible.
- Check a sample for bacterial contamination in
  a) Saboraud fluid medium at 31°C for 1 week (see Appendix 4),
  b) brain heart infusion broth at 37°C for 1 week (see Appendix 4).
- Store at room temperature.

TABLE 4

Glucose (10% w/v)

- Dissolve 100 g glucose in distilled water.
- Make up to 1 litre with distilled water.
- Sterilise by filtration using a 0.22 μm membrane filter.
- Bottle in 5 and 10 ml amounts and store at -20°C.

TABLE 5

Hepes buffer

- Dissolve 47.6 g Hepes in 200 ml distilled water and adjust to pH 8.1 with NaOH.
- Sterilise by filtration using a 0.22 μm membrane filter and store as aliquots of 5 ml at room temperature.
- Check for bacterial contamination in
  a) Saboraud fluid medium at 31°C for 1 week (Appendix 4),
  b) brain heart infusion broth at 37°C for 1 week (Appendix 4).

TABLE 6

Antibiotic solution × 100

Na Benzyl penicillin G (Crystopen)	10,000,000 units
Streptomycin sulphate	10 g
Distilled water to	1 litre

- Sterilise by filtration using a 0.22 μm membrane filter.
- Bottle in 5, 20, and 50 ml amounts.
- Check for bacterial contamination in
   a) Saboraud fluid medium at 31°C for 1 week (Appendix 4),
   b) brain heart infusion broth at 37°C for 1 week (Appendix 4).
- Store at −20°C.

The penicillin and streptomycin can be obtained from Glaxo Labs. (see Appendix 3).

TABLE 7

Phenol red 1% (for use as an indicator in BSS and media)

- Dissolve 10 g phenol red in 245 ml of distilled water with 5 ml of 5 N NaOH.
- Add N.HCl drop by drop till a deep blood red colour is obtained.
- Make up to 1 litre with distilled water.
- Filter once through Whatman's No. 1 filter paper.
- Bottle and store at 4°C.

TABLE 8

Eagle's non-essential amino acids (NEAA) × 100

	g/l
L-Alanine	0.89
L-Asparagine–H$_2$O	1.50
L-Aspartic acid	1.33
L-Glutamic acid	1.47
Glycine	0.75
L-Proline	1.15
L-Serine	1.05

TABLE 9

Lactalbumin hydrolysate purchased as a dry powder from Nutritional Biochemicals (U.S.A.) is kept in brown sealed bottles. It is dissolved to 5% in Hanks' BSS and autoclaved at 115°C for 10 min when it may be kept at room temperature for a month or more. If kept frozen a precipitate forms which will redissolve on heating in a boiling water bath. For use it is diluted tenfold to 0.5% with growth medium.

TABLE 10

Chick embryo extract

- Aseptically remove from the egg 10-day-old chick embryos as described on p. 79 and wash them in Hanks' BSS. Homogenise in a blender for 60 s at maximum speed in an equal volume of Hanks' BSS.
- Stand at 4°C for 1 h and centrifuge at 35,000 g for 20 min.
- Freeze supernatant at −20°C overnight. Then thaw, re-spin and re-freeze in aliquots for storage.

TABLE 11

Tryptose phosphate broth

Tryptose phosphate broth
(Difco bacto or Oxoid – Appendix 3)   147.5 g
Distilled water to   5   litres

- Dispense in 50 ml amounts in 4 oz medical flat bottles.
- Autoclave at 15 lb pressure for 15 min to sterilise.
- Keep at 37°C for 7 days.
- Check each bottle for turbidity before storing at room temperature.

TABLE 12

Eagles' media formulations (amino acid and vitamin components)

	BME	MEM	Glasgow MEM	Dulbecco's MEM
Amino acids (mg/l)				
L-arginine–HCl		126.0	42.0	84.0
L-arginine	17.4			
L-cystine	12.0	24.0	24.0	48.0
L-glutamine	292.0	292.0	292.0	584.0
L-glycine				30.0
L-histidine HCl. H_2O		42.0	21.0	42.0
L-histidine	8.0			
L-isoleucine	26.0	52.0	52.4	105.0
L-leucine	26.0	52.0	52.4	105.0
L-lysine–HCl		72.5	73.1	146.0
L-lysine	29.2			
L-methionine	7.5	15.0	15.0	30.0
L-phenylalanine	16.5	32.0	33.0	66.0
L-serine				42.0
L-threonine	24.0	48.0	47.6	95.0
L-tryptophan	4.0	10.0	8.0	16.0
L-tyrosine	18.0	36.0	36.2	72.0
L-valine	23.5	46.0	46.8	94.0
Vitamins (mg/l)				
biotin	1.0			
D-Ca pantothenate	1.0	1.0	1.0	4.0
choline chloride	1.0	1.0	1.0	4.0
folic acid	1.0	1.0	1.0	4.0
i-inositol	2.0	2.0	2.0	7.2
nicotinamide	1.0	1.0	1.0	4.0
pyridoxal–HCl	1.0	1.0	1.0	4.0
riboflavin	0.1	0.1	0.1	0.4
thiamin–HCl	1.0	1.0	1.0	4.0

The amino acids and vitamins are made up in either Earle's or Hanks' BSS. Dulbecco's modification is made up in Earle's BSS containing extra bicarbonate (3.7 g/l) and glucose (4.5 g/l) and supplemented with ferric nitrate (0.1 mg Fe(NO$_3$)$_3$.9H$_2$O/l). The formulations are based on those originally prescribed by Eagle (1955a, 1959).

TABLE 13

Concentrated (× 10) Eagle's MEM stock solution

Commercial suppliers, e.g. Flow Laboratories or Gibco-Biocult (Appendix 3), sell Eagle's MEM amino acids as a × 50 stock and the vitamins as × 100 stock. The former may be stored at room temperature but the latter should be stored frozen until required.

They may be combined aseptically, but as glucose and glutamine also have to be added and these require filter sterilising, it is more convenient to sterilise after preparation of a 10 × stock:

MEM* amino acids × 50	200	ml
MEM vitamins × 100	100	ml
L-glutamine**	2.925	g
glucose	45	g
glass distilled water to	1	litre

– Adjust the pH to 7.1 using 5 N NaOH (approx. 10 ml).
– Sterilise by filtration using a 0.22 μm membrane filter.
– Bottle in 50, 100, 200, and 400 ml amounts.
– Check for bacterial contamination in
   a) Saboraud fluid medium at 31°C for 1 week (Appendix 4),
   b) brain heart infusion broth at 37°C for 1 week (Appendix 4).
– Store at 4°C until the results of the checks are known.

* To make Glasgow MEM use 200 ml BME amino acids × 100.
** L-Glutamine may be omitted and added to 1 × medium immediately before use (1 ml of 200 mM solution per 100 ml medium).

TABLE 14

Concentrated (× 10) Dulbecco's MEM stock solution

This is made up similarly to Eagle's MEM × 10 stock solution but using Dulbecco's amino acids × 25.

to make 1 litre

Dulbecco's amino acids × 25	400 ml
MEM vitamins × 100	400 ml
glucose	45 g
distilled water	200 ml
pH to 7.1 using 5 N NaOH	

- Sterilise by filtration using a 0.22 μm membrane filter.
- Bottle in 50 and 100 ml amounts.
- Check for bacterial contamination as above.

N.B. L-Glutamine is added to 1 × medium immediately before use (10 ml 200 mM glutamine to 500 ml growth medium).

TABLE 15

Preparation of Eagle's media from concentrated stock solutions

The following formulations are based on the use of Eagle's MEM 10 × made up without salts (see Table 13) or Dulbecco's MEM × 10 (see Table 14). However, if the 10 × stock is made with salts (i.e. if it is made from powder) then the BSS should be replaced by sterile glass-distilled water.

(a) *Eagle's minimum essential medium supplemented with 10% calf serum* (EC10) (used as growth medium for most routine cell lines, e.g. L929, HeLa, CHO – supplemented with proline for pro⁻ lines - BHK21/C13 and some polyoma transformed lines).

BSS*	450 ml
Eagle's MEM 10 × stock	50 ml
NaHCO$_3$ (5.6% - Table 3)	20 ml
Calf serum	50 ml
Antibiotics (× 100 – Table 6)	5 ml

(b) *Glasgow modification of Eagle's medium supplemented with 10% calf serum and 10% tryptose phosphate broth* gives improved growth of BHK21/C13 and polyoma transformed lines.

BSS*	400 ml
Glasgow MEM 10 × stock	50 ml
NaHCO$_3$ (5.6% – Table 3)	20 ml
Calf serum	50 ml
Tryptose phosphate (Table 11)	50 ml
Antibiotics (×100 – Table 6)	5 ml

(c) *Eagle's minimum essential medium supplemented with 10% foetal calf serum and non-essential amino acids* is recommended for growth of BSCl, CVl and PT-K cells and may produce better growth of L929 and HeLa cells.

BSS*	450 ml
Eagle's 10 × stock	50 ml
NaHCO$_3$ (5.6% – Table 3)	20 ml
Foetal calf serum	50 ml
NEAA (×100 – Table 8)	5 ml
Antibiotics (×100 – Table 6)	5 ml

(d) *Dulbecco's MEM* may be supplemented with 10% calf serum, 20% foetal calf serum or mixtures of the two.

BSS*	400 ml
Dulbecco's 10 × stock	50 ml
NaHCO$_3$ (5.6% – Table 3)	30 ml
Foetal calf serum	100 ml
Antibiotics (×100 – Table 6)	5 ml
Glutamine (200 mM)	10 ml

* The BSS in all cases is made up from 50 ml BSS 10 × stock diluted to the appropriate volume with distilled water and sterilised by autoclaving (Table 1).

N.B. All ingredients should be added aseptically to the bottle of BSS and the medium warmed to 37°C before use. The media may be stored at 4°C for a period of 2–3 weeks.

If antibiotics are omitted (the recommended procedure) a sample of the growth medium should be incubated at 37°C for 3 days and at room temperature for a further two days and bottles showing contamination discarded.

TABLE 16

Ham's medium F12

From a formulation of Ham (1965)

	mg/l		mg/l
Amino acids		L-cysteine–HCl.H$_2$O	35.12
L-alanine	3.91	L-glutamic acid	14.71
L-arginine–HCl	210.7	L-glutamine	146.2
L-asparagine	15.01	glycine	7.51
L-aspartic acid	13.31	L-isoleucine	3.94

	mg/l		mg/l
L-leucine	13.12	vitamin B12	1.357
L-lysine–HCl	36.54	Salts	
L-methionine	4.48	KCl	223.65
L-phenylalanine	4.96	NaCl	7.6
L-proline	34.53	$Na_2HPO_4.7H_2O$	268.1
L-serine	10.51	$FeSO_4.7H_2O$	0.834
L-threonine	11.91	$MgCl_2.6H_2O$	122.0
L-tryptophan	2.04	$CaCl_2.2H_2O$	44.11
L-tyrosine	5.44	$CuSO_4.5H_2O$	0.0025
L-valine	11.71	$ZnSO_4.7H_2O$	0.863
Vitamins		$NaHCO_3$	1.176
biotin	0.007	Other components	
Ca-pantothenate	0.258	phenol red	1.242
choline chloride	13.96	glucose	1801.6
folic acid	1.324	Na-pyruvate	110.1
lipoic acid	0.206	putrescine–$2HCl$	0.161
niacinamide	0.037	hypoxanthine	4.083
pyridoxine–HCl	0.062	myo-inositol	18.02
riboflavin	0.038	thymidine	0.727
thiamin–HCl	0.337	linoleic acid	0.084

TABLE 17

McCoy's 5a medium (RPMI 1629)

Originally formulated by McCoy et al. (1959) and modified by Hsu and Kellogg (1960) and Iwakata and Grace (1964)

	mg/l		mg/l
Amino acids		L-hydroxyproline	19.70
L-alanine	13.90	L-isoleucine	39.36
L-arginine–HCl	42.10	L-leucine	39.36
L-asparagine	45.00	L-lysine–HCl	36.50
L-aspartic acid	19.97	L-methionine	14.90
L-cysteine	31.50	L-phenylalanine	16.50
L-glutamic acid	22.10	L-proline	17.30
L-glutamine	219.20	L-serine	26.30
glycine	7.50	L-threonine	17.90
L-histidine–HCl–H_2O	20.96	L-tryptophan	3.10

	mg/l		mg/l
L-tyrosine	18.10	thiamin–HCl	0.20
L-valine	17.60	vitamin B12	2.00
Vitamins		Salts	
ascorbic acid	0.50	$CaCl_2$	100.0
biotin	0.20	KCl	400.0
choline chloride	5.00	$MgSO_4.7H_2O$	200.0
D-Ca-pantothenate	0.20	NaCl	6460.0
folic acid	10.00	$NaHCO_3$	2200.0
i-inositol	36.00	$NaH_2PO_4.H_2O$	580.0
nicotinamide	0.50	Other components	
nicotinic acid	0.50	Phenol red	10.0
p-aminobenzoic acid	1.00	glucose	3000.0
pyridoxal–HCl	0.50	glutathione	0.50
pyridoxine–HCl	0.50	bacto-peptone	600.0
riboflavin	0.20	foetal calf serum	0–30%

TABLE 18

Medium 199

As formulated by Morgan et al. (1950, 1955)

	mg/l		mg/l
Amino acids		DL-methionine	30.0
L-alanine	50.0	DL-phenylalanine	50.0
L-arginine–HCl	70.0	L-proline	40.0
DL-aspartic acid	60.0	DL-serine	50.0
L-cysteine–HCl	0.1	DL-threonine	60.0
L-cystine	20.0	DL-tryptophan	20.0
DL-glutamic acid–H_2O	150.0	L-tyrosine	40.0
L-glutamine	100.0	DL-valine	50.0
glycine	50.0	Vitamins	
L-histidine–HCl	20.0	ascorbic acid	0.050
L-hydroxyproline	10.0	d-biotin	0.010
DL-isoleucine	40.0	calciferol	0.100
DL-leucine	120.0	Ca-pantothenate	0.010
L-lysine–HCl	70.0	choline chloride	0.500

	mg/l		mg/l
folic acid	0.010	pyridoxal–HCl	0.025
i-inositol	0.050	pyridoxine–HCl	0.025
menadione	0.010	riboflavin	0.010
niacin	0.025	thiamine–HCl	0.010
niacinamide	0.025	vitamin A (acetate)	0.14
para-aminobenzoic acid	0.050		

Inorganic salts	Hanks' salts (mg/l)	Earle's salts (mg/l)
$CaCl_2$ (anhyd)	140.0	200.0
ferric nitrate $Fe(NO_3)_3$	0.1	0.1
KCl	400.0	400.0
KH_2PO_4	60.0	–
$MgSO_4.7H_2O$	200.0	200.0
NaCl	8000.0	6800.0
$NaHCO_2$	350.0	2200.0
$NaH_2PO_4.H_2O$	–	140.0
$Na_2HPO_4.2H_2O$	60.0	–

Other components	mg/l		mg/l
adenine sulphate	10.00	glutathione	0.05
adenosinetriphosphate		guanine–HCl	0.30
(disodium salt)	1.00	hypoxanthine	0.30
adenylic acid	0.20	phenol red	20.0
alpha tocopherol		ribose	0.50
phosphate (sodium		sodium acetate	50.00
salt)	0.01	thymine	0.30
cholesterol	0.20	Tween 80c	20.00
deoxyribose	0.50	uracil	0.30
glucose	1000.0	xanthine	0.30

TABLE 19

Medium NCTC 135

As formulated by Evans et al. (1964)

	mg/l		mg/l
Amino acids		menaphthone sodium	
L-alanine	31.48	bisulphite trihydrate	0.048
L-α-amino-*n*-butyric acid	5.51	nicotinic acid	0.0625
L-arginine–HCl	31.16	nicotinamide	0.0625
L-asparagine–H$_2$O	9.19	*p*-aminobenzoic acid	0.125
L-aspartic acid	9.91	pyridoxal–HCl	0.0625
L-cystine, disodium salt	12.41	pyridoxine–HCl	0.0625
L-glutamic acid	8.26	riboflavin	0.025
L-glutamine	135.7	thiamin–HCl	0.025
glutathione	10.00	DL-α-tocopherol phos-	
glycine	13.51	phate, disodium salt	0.025
L-histidine–HCl . H$_2$O	26.65	vitamin A acetate	0.29
L-hydroxyproline	4.09	vitamin B12	10.00
L-isoleucine	18.04	Salts	
L-leucine	20.44	CaCl$_2$.2H$_2$O	264.9
L-lysine–HCl	38.43	KCl	400.0
L-methionine	4.44	MgSO$_4$.7H$_2$O	204.8
L-ornithine HCl	9.41	NaCl	6800
L-phenylalanine	16.53	NaHCO$_3$	2200
L-proline	6.13	NaH$_2$PO$_4$2H$_2$O	158.3
L-serine	10.75	Other components	
taurine	4.18	cocarboxylase	1.00
L-threonine	18.93	coenzyme-A	2.50
L-tryptophan	17.5	deoxyadenosine–H$_2$O	10.00
L-tyrosine	16.44	deoxycytidine–HCl	10.00
L-valine	25.00	deoxyguanosine–H$_2$O	10.00
Vitamins		FAD, disodium salt	1.00
L-ascorbic acid	50.00	D-glucosamine–HCl	3.85
biotin	0.025	glucose	1000
choline chloride	1.25	D-glucuronolactone	1.80
calciferol	0.25	5-methylcytosine	0.10
D-calcium pantothenate	0.025	NAD	7.00
folic acid	0.025	NADP, sodium salt	
i-inositol	0.125	dihydrate	1.00

	mg/l		mg/l
sodium acetate	30.14	Tween 80	12.50
sodium glucuronate . H_2O	1.80	UTP, trisodium salt	
sodium phenol red	20.00	dihydrate	1.00
thymidine	10.00		

TABLE 20

Medium CMRL 1066

As formulated by Parker et al. (1957).

	mg/l		mg/l
Amino acids		cholesterol	0.200
L-alanine	25.0	choline chloride	0.500
L-arginine–HCl	70.0	folic acid	0.010
L-aspartic acid	30.0	*i*-inositol	0.050
L-cysteine–HCl . H_2O	260.0	niacin	0.250
L-cystine	20.0	niacinamide	0.025
L-glutamic acid	75.0	para-aminobenzoic acid	0.050
L-glutamine	100.0	pyridoxal–HCl	0.025
glycine	50.0	pyridoxine–HCl	0.025
L-histidine–HCl . H_2O	20.0	riboflavin	0.010
hydroxy-L-proline	10.0	thiamine–HCl	0.010
L-isoleucine	20.0	Inorganic salts	
L-leucine	60.0	$CaCl_2$ (anhyd)	200.0
L-lysine–HCl	70.0	KCl	400.0
L-methionine	15.0	$MgSO_4.7H_2O$	200.0
L-phenylalanine	25.0	NaCl	6799.0
L-proline	40.0	$NaHCO_3$	2200.0
L-serine	25.0	$NaH_2PO_4.H_2O$	140.0
L-threonine	30.0	Other components	
L-tryptophan	10.0	cocarboxylase	1.0
L-tyrosine	40.0	coenzyme-A	2.5
L-valine	25.0	deoxyadenosine	10.0
Vitamins		deoxycytidine	10.0
ascorbic acid	50.000	deoxyguanosine	10.0
biotin	0.010	diphosphopyridine	
Ca-pantothenate	0.010	nucleotide	7.0

	mg/l		mg/l
ethanol	16.0	sodium acetate . $3H_2O$	83.0
flavin adenine		sodium glucuronate . H_2O	4.2
dinucleotide	1.0	thymidine	10.0
glucose	1000.0	triphosphopyridine	
glutathione	10.0	nucleotide	1.0
5-methyl-		Tween 80	5.0
deoxycytidine	0.1	uridine triphosphate	1.0
phenol red	20.0		

APPENDIX 2

Stains and fixatives

A number of stains and fixatives have been referred to throughout the book and these can be located by using the index. Here are listed some of the more common ones.

1. Acetic orcein
– Fix cells in methanol : acetic acid (3 : 1) and air-dry.
– Stain for 3–5 min with 2% orcein in 4% acetic acid.
– Rinse in 40% methanol and air-dry.

2. Acridine orange fluorescence stain
   To a suspension of cells in BSS is added a drop of an aqueous 0.01% solution of acridine orange. On examination under the fluorescent microscope nuclei appear yellowish green and cytoplasm red.

3. Carnoy's fixative
   1 part glacial acetic acid
   3 parts absolute ethanol

4. Formal saline
– Dissolve 5 g NaCl and 15 g Na_2SO_4 in distilled water and make up to 900 ml.
– Filter 40% formaldehyde using Green's filter paper No. $904\frac{1}{2}$ and add 100 ml to the salt solution.

5. Giemsa stain
– To 30 g Giemsa add 1,980 ml glycerol.
– Heat at 56°C for 90–120 min.
– Add 1,980 ml methanol and mix well.
– Stand at room temperature for 7 days and filter through Green's $904\frac{1}{2}$ filter paper.
For use : Dilute 1/10 in buffered distilled water or use undiluted.

6. Gram stain
a. Methyl violet 6B   8   g
   Absolute ethanol   80   ml
   1% aqueous ammonium oxalate   320   ml
b. Iodine   1   g
   Potassium iodide   2   g
   Distilled water   100   ml
c. Safranin   2.5 g
   Ethanol   100   ml
   Distilled water to   1   l
- Pour methyl violet solution (a) onto the fixed preparation and remove after 30 sec.
- Wash in tap water.
- Pour on iodine solution (b) and remove after 1 min.
- Wash with water and decolorise with ethanol until no more stain is removed.
- Apply safranin counter stain solution (c) for 2 min.
- Wash with water and dry at room temperature.
- Examine under oil immersion.

7. Haematoxylin and eosin
The haematoxylin is dissolved in 95% ethanol and to 6 ml of a 17% solution is added 100 ml saturated ammonium alum solution and 0.5 g mercuric oxide. After boiling and cooling 25 ml of glycerol and 25 ml of methanol are added and the whole mixture filtered.
- Stain the fixed cells overnight in a 100-fold dilution of haematoxylin mixture.
- Rinse in water for 15 min.
- Counter stain in 0.5% eosin Y for 0.5–1 min and rinse.
- Dehydrate in 95% and absolute ethanol.
Nuclei are stained blue and cytoplasm red.

8. Leishman stain
- To 7.5 g Leishman stain add 5 l methanol and shake at intervals over 5 days before use.

9. May–Grunwald Giemsa stain
- Dilute 9 ml stock Giemsa (0.3% in glycerol, methanol) with 90 ml 0.1 M phosphate buffer pH 6.8 and add 1ml of May–Grunwald stain (0.25% w/v in methanol).

10. Methylene blue
The stain is 30 ml saturated ethanolic solution of methylene blue mixed with 100 ml 0.01% aqueous KOH.
- Stain the preparation for 3 min and wash with water.
- Blot dry and examine microscopically under oil immersion.
This stain is used so that all bacteria are more easily seen under high power objectives.

11. Orcein stain

A 2% solution of natural orcein (Gurr's) in 60% glacial acetic acid is prepared by dissolving the orcein in boiling glacial acetic acid with stirring. Cool to 55°C. Add distilled water to make the acid 60%. Cool to room temperature. Filter twice through Whatman filter paper No. 1.

APPENDIX 3

Suppliers

The following suppliers are those with whom I have had contact. The list is not exclusive, nor does the presence of the name of a supplier imply recommendation of their products. The suppliers, or their agents, are largely British and in some instances there is a Scottish bias and this again is largely a result of my experience. Further information may be obtained from the Laboratory Equipment Directory published annually by Morgan-Grampian Book Publishing Co. Ltd., 30 Calderwood St., London SE15, 6QH.

Abbot Laboratories, North Chicago, IL, U.S.A.

American Type Tissue Collection, Rockville, MD 20852, U.S.A.

Amicon Ltd., 57 Queen's Road, High Wycombe, Bucks.

Anderman & Co. Ltd., Central Avenue, East Molesey, Surrey KT8 OQZ.

Charles Austen Pumps Ltd., 100 Royston Road, Byfleet, Surrey.

Becton Dickinson U.K. Ltd., (Falcon Plastics) York House, Empire Way, Wembley, Middlesex HA9 OPS.

Bellco, *see* A.R. Horwell Ltd.

A. & J. Beveridge Ltd., 5 Bonnington Road Lane, Edinburgh EH6 5BP.

Bioassay Systems (L.H. Engineering Co. Ltd.) Bells Hill, Stoke Poges, Bucks. SL2 4EG.

Bio-Rad Laboratories Ltd., Caxton Way, Holywell Industrial Estate, Watford, Herts. WD1 8RP.

British Drug Houses Ltd., Poole, Dorset BH12 4NN.

British Oxygen Co., 150 Polmadie Road, Glasgow G5.

A. Browne Ltd., Chancery House, Abbey Gate, Leicester LE4 OAA.

Calbiochem Ltd., 79/81 South Street, Bishops Stortford, Herts. CM23 3AL.

Connaught Laboratories, *see* A.R. Horwell Ltd.

Coulter Electronics Ltd., Coldharbour Lane, Harpenden, Herts. AL5 4UN.

Corning Ltd., Halstead, Essex CO9 2DX.

Decon Laboratories Ltd., Ellen Street, Portslade, Brighton BN4 1EQ.

Dynatech Labs. Ltd., *see* Gibco-Biocult.

Difco Laboratories, P.O. Box No. 14B, Central Avenue, East Molesey, Surrey KT8 OSE.

Distillers Co. Ltd., Port Dundas Distillery, 74 North Canal Bank Street, Glasgow G4.

Dow Corning Co., *see* McFarlane Robson.

Durham Chemical Distributors, Birtley, Tyne and Weir.

Elga Group, Lane End, High Wycombe, Bucks.

Epsom Glass Industries Ltd., *see* McFarlane Robson.

Evans Medical Supplies Ltd., Ruislip, London.

Falcon Plastics, *see* Becton Dickinson U.K. Ltd.

Fisons Scientific Apparatus Ltd., Loughborough, Leics. LE11 ORG.

Flow Laboratories, Victoria Park, Heatherhouse Road, Irvine, Ayrshire.

Forma CO_2 Incubators–Raven Scientific Ltd., P.O. Box 2, Haverhill, Suffolk.

A. Gallenkamp & Co. Ltd., Braeview Place, Nerston, East Kilbride, Glasgow G74 3XJ.

Gelman Hawksley & Sons Ltd., Harrenden Road, Brackmills, Northampton.

Gibco-Biocult Labs. Ltd., 3 Washington Road, Sandyford Industrial Estate, Paisley PA3 4EP.

Glaxo Laboratories Ltd., Greenford, Middlesex.

Grant Instruments Ltd., Barrington, Cambridge CB2 5QZ.

Green's Filter Paper, *see* Whatman Lab. Sales Ltd.

G.T. Gurr Ltd., *see* Searle Scientific Services.

W.C. Haraeus GmbH, Postfach 169, D6450 Hanau.

Hotpack Ltd., *see* Gibco-Biocult.

Hopkin & Williams, P.O. Box 1, Romford, RM1 1HA.

Hoechst U.K. Ltd., Hoechst House, Salisbury Road, Hounslow, Middlesex TW4 6JH.

A.R. Horwell (Bellco) 2, Grangeway, Kilburn High Road, London NW6 2BP.

V.A. Howe & Co. Ltd., 88, Peterborough Road, London SW6 3EP.

Hyflo Pumps, *see* Metcalf Bros. Ltd.

Ilford Ltd.–Hamilton Tait Ltd., Polmadie Industrial Estate, Toryglen Street, Glasgow G5 OBH. or Ilford, London.

Kodak Ltd. (Northern Region) Dallimore Road, Wythenshawe, Manchester M23 9NJ, or Acornfield Road, Kirby, Liverpool L33 7UX.

Linbro, *see* Flow Laboratories.

3M Co. Ltd., *see* McFarlane Robson.

LEEC–Private Road No. 7, Colwick Estates, Nottingham NG4 2AJ.

Luckham Ltd., Labro Works, Victoria Gardens, Burgess Hill, Sussex RH15 9QV.

May & Baker Ltd. Dagenham, Essex.

Millipore (U.K) Ltd., Millipore House, Abbey Road, Park Royal, London NW10 7SP.

McFarlane Robson Ltd., Burnfield Avenue, Thornliebank, Glasgow G46 7TP.
  (A local supplier. Other districts will have their own local supplier).

Metcalf Bros. Ltd., Cranbourne Road, Potters Bar, Herts.

New Brunswick Scientific (U.K.) Ltd., 40 Wellington Street, London WC2E 7BD.

Nutritional Biochemical Corp., *see* Uniscience Ltd.

Nunc, *see* Gibco-Biocult.

Olympus Microscopes, *see* Gallenkamp

Oxoid Ltd., Wade Road, Basingstoke, Hants. RG24 OPW.

Petriperm, see W.C. Haraeus GmbH.

Pharmacia (G.B.) Ltd., Paramount House, 75 Uxbridge Road, Ealing, London W5 58S.

W.R. Prior & Co. Ltd., London Road, Bishop's Stortford, Herts.

Quickfit & Quartz Ltd., see McFarlane Robson.

Repelcote, see Hopkin & Williams.

Sartorious Ltd., see V.A. Howe.

Searle Scientific Services, Coronation Road, Cressex, High Wycombe, Bucks.

Sigma London Chem. Co. Ltd., Fancy Road, Poole, Dorset BH17 7NH.

Schleicher and Schuell Ltd., see Anderman & Co. Ltd.

Techmation Ltd., 58 Edgware Way, Edgware HA8 8JP.

Titertek, see Flow Laboratories.

Uniscience Ltd., Uniscience House, 8 Jesus Lange, Cambridge CB5 8BA.

Union Carbide (U.K.) Ltd., Redworth Way, Aycliffe Industrial Est. Aycliffe, Co. Durham.

Vir Tis Co. Inc., Gardiner, New York 12525, see Techmation Ltd.

Voss Instruments Ltd., Faraday Works, High Street, Malden, Essex CM9 7EY.

Whatman Labsales Ltd. (Reeve Angel), c/o East Anglia Chemicals Ltd. Lady Lane Industrial Estate, Hadleigh, Ipswich 1P7 6BQ.

Wild Heerbrugg Ltd., CH-9435 Heerbrugg, Switzerland.

R. & J. Wood Ltd., 39 Back Sneddon St., Paisley PA3 2DE, Scotland.

APPENDIX 4

Sterility checks

1. Beef heart infusion broth
- Add 50 g beef heart infusion (Difco Labs., Detroit) to 900 ml distilled water at 50°C. Stand for 1 h.
- Bring slowly to the boil.
- Filter through a double layer of Whatman No. 12 filter paper.
- Distribute into bottles and sterilise by autoclaving for 15 min at 15 lb pressure.

2. Brain heart infusion broth
- Dissolve 37.0 g of brain heart infusion broth powder (Oxoid, Basingstoke) in 1 l of distilled water.
- Bottle in 25 ml and 40 ml amounts in 2 oz medical flat bottles.
- Autoclave at 15 lb pressure for 15 min to sterilise.
- Store at room temperature.

3. Saboraud fluid medium
- Dissolve 30 g of Saboraud fluid medium base (Oxoid, Appendix 3) in 1 l of distilled water.
- Bottle in 25 ml and 40 ml amounts in 2 oz medical flat bottles.
- Autoclave at 15 lb pressure for 15 min to sterilise.
- Store at room temperature.

4. Blood agar plate
- Dissolve 40 g blood agar base (Oxoid, Basingstoke) in 1 l of distilled water by standing in a boiling water bath.
- Aliquot in 50 ml parts and autoclave at 121°C for 15 min.
- To 50 ml agar base (heated to melt and cooled to 45°C) add 5 ml horse blood.
- Pour 5 ml into 5 cm Petri dishes and allow to cool.
The horse blood can be delivered regularly, e.g. every two weeks, from the suppliers (Oxoid Ltd., Basingstoke).

5. PPLO Agar

A. PPLO agar base
- To make 1 l PPLO Agar base mix 34.0 g Bacto-PPLO Agar (Difco Labs.) with 1 l cold distilled water in a flask and stand this in a bucket of boiling water until the agar dissolves.
- While molten, dispense in 25 ml amounts in 4 oz medical flat bottles.
- Autoclave at 15 lb pressure for 15 min to sterilise.
- Store at room temperature.

B. PPLO agar plates	for 10 plates (ml)
PPLO agar base	25
Horse serum	7.5
Yeast extract (see below)	3.75
Trypstose phosphate broth (Difco)	3.75
Thallous acetate (1.25%)	0.4
Penicillin	0.15

- Boil the agar (in a water bath) until molten and hold at 45°C.
- Heat the serum and yeast extract to 45°C.
- Add the horse serum, yeast extract, tryptose phosphate broth, thallous acetate, and penicillin to the agar.
- Mix well and dispense 4 ml amounts into 50 mm Petri dishes.
- Allow to solidify and store at 4°C.

C. Yeast extract
- Add 250 g dried baker's yeast to 1 l distilled water and bring to the boil.
- Filter through a double layer of Whatman No.12 filter paper.
- Add NaOH to raise the pH to 8.0 and dispense into 10 ml aliquots, autoclave and store at −20°C.

6. *Soy peptone–yeast dialysate medium* (Kenny, 1973)

A. Basic broth

Soy peptone (Sheffield Chem. Co. Norwich, NY)	30 g
NaCl	5 g
H_2O	1 l

Adjust to pH 7.4 and add 10 g of Difco Noble Agar if a basic agar is required. Autoclave and store at room temperature.

B. Yeast dialysate
450 g Fleischmann's active dried yeast is added to 1250 ml H_2O and autoclaved for 5 min at 121°C. It is then dialysed against 1 l of water at 4°C for 2 days. The dialysate is autoclaved and stored frozen.

C. Complete medium

Basic broth (or agar)	65 ml
Yeast dialysate	10 ml
Horse serum	25 ml
Penicillin	20,000 units
Thallium acetate (3.3%)	1 ml

Inclusion of arginine (16 mM) and 0.4 mg% phenol red indicates the presence of arginine deiminase by formation of alkali (purple colouration).

APPENDIX 5

Assays

1. DNA, RNA and protein estimations

Using the following procedure DNA, RNA and protein may all be conveniently assayed on the same batch of cells.
a. Harvest cells by trypsinisation and count them using a haemocytometer (§ 8.2.) or electronic cell counter (§ 8.2.2).

b. Wash the cell suspension twice in PBS to remove extraneous protein followed by two washes in 5% trichloracetic acid to remove acid soluble material.

c. Dissolve about 10^7 cells in 1.0 ml 0.10 N NaOH (heat to 40°C if necessary).

d. Take 0.15 ml solution plus 2.85 ml 0.7 N NaOH. Use for protein estimations (see below).

e. To the remaining 0.85 ml solution add 0.15 ml 3.3 N PCA. Incubate at 70°C for 30 min. Centrifuge and retain the supernatant (hot acid extract) for estimation of DNA and RNA.

A. DNA estimation

The diphenylamine reagent is made up according to Burton (1956). To 500 ml glacial acetic acid add 7.5 ml concentrated sulphuric acid and 7.5 g diphenylamine. Mix cautiously and store in a brown bottle at room temperature. Just before use add 0.1 ml 1.6% acetaldehyde to 20 ml of the above mixture.

– Place 0.4 ml aliquots of the hot acid extracts (from 1:e above) in tubes and make up to 1.0 ml with 0.5 N perchloric acid.

– Add 2.0 ml diphenylamine reagent and mix.

– Cover and stand at 30°C for 18–48 h and read the absorbance at 600 nm.

A standard DNA solution can be made by dissolving native DNA in 50 mM KCl and from its spectrum its concentration may be calculated (Hirschman and Felsenfeld, 1966). Calf thymus DNA dissolved at 1 mg/ml has an E_{260} of 18.5. The standard DNA solution should then be made 0.5 N with respect to perchloric acid and heated at 70°C for 30 min to dissolve the DNA. This solution should be diluted to 100 μg/ml with 0.5 N perchloric acid and various amounts up to 1.0 ml used as described above to prepare a standard curve.

B. RNA estimation

A standard RNA solution is prepared by dissolving RNA in 0.05 M NaOH at 50 μg/ml.

– Place 0.2 ml aliquots of the hot acid extracts from 1:e above in large test tubes (6 × 5/8") and dilute to 1.5 ml with water (for the standard curve use up to 1.5 ml of the standard RNA solution).

– Add 1.5 ml of 0.03% FeCl$_3$ in concentrated HCl and 0.1 ml 20% freshly prepared orcinol in 95% ethanol.

– Mix thoroughly and place in a *vigorously* boiling water bath for 30 min. Use glass 'dew-drops' to prevent loss of volume.

– Cool in ice-water slurry and read the absorbance at 665 nm.

C. Protein estimation

This is a modification of the method of Lowry et al. (1951). The Folin and Ciocalteau's Reagent is available from British Drug Houses Ltd. (Appendix 3) and should be adjusted to 2 N before use. Solution A is prepared daily by adding to

100 ml 13% Na_2CO_3, 3 ml 4% NaK tartrate ($NaKC_4H_4O_64H_2O$; Rochelle salt) and 3 ml 2% $CuSO_4$. $5H_2O$ in this order and mixing immediately. As standard, bovine serum albumin (150 $\mu g/ml$) is used.

- Place 0.2 and 0.6 ml aliquots of the cell digests from 1 : d above into tubes and make up to 1.5 ml with 0.66 N NaOH.
- Add 1.5 ml Solution A and mix.
- After *exactly* 10 min add 0.5 ml Folin–Ciocalteau reagent from an 0.5 ml 'blow-out' pipette and *mix thoroughly and immediately*.
- Stand at room temperature for 30 min, and read the absorbance at 625 nm.

2. Fluorescent DNA assays

The diphenylamine method can be scaled down to increase sensitivity but fluorescent methods are available using diaminobenzoic acid (Kissane and Robins, 1958; Hinegardner, 1971) or ethidium bromide (Klotz and Zimm, 1972) which are sensitive down to 0.1 μg DNA. In Hinegardner's method a salt-free DNA sample is dried down in a siliconised tube and then heated at 60°C for 45 min with 0.1 ml repurified diaminobenzoic acid (0.4 g/ml water). 1.5 ml 1.0 N HCl is then added and the fluorescence measured using an exciting wavelength of 410 nm (405 primary filter) and an emission wavelength of 500 nm (combination 2A-12, 65A secondary filter). This method may be applied directly to cells. After counting, the cells should be fixed in 10% neutral formalin buffered to pH 8 with 0.1 M borate and then washed several times with water before drying down. If several different volumes of the same suspension are assayed then the slope of the line (volume vs. emission) gives a measure of the DNA concentration.

APPENDIX 6

Abbreviations

ACTH	adrenocorticotrophic hormone
Ad2	adeno 2 virus
AMP, ADP, ATP	adenosine monophosphate, diphosphate, triphosphate
ATCC	American Type Culture Collection
BHK cells	Baby hamster kidney cells
BME	basal medium (Eagle)
BPL	β-propiolactone
BSS	balanced or buffered salt solution
BUdr	bromodeoxyuridine
CHO cells	Chinese hamster ovary cells

CMF	calcium and magnesium free BSS
CMP, CDP, CTP	cytidine monophosphate, diphosphate, triphosphate
CMRL	Connaught Medical Research Laboratories
CMS	Calgon metasilicate
CPE	cytopathic effect
dAMP, dTMP	deoxyadenosine monophosphate, deoxythymidine monophosphate, etc.
Decon	a detergent
DePex	mounting medium available from G.T. Gurr Ltd. (Appendix 3)
EC10	Eagle's MEM supplemented with 10% calf serum
EDTA	ethylene diamine tetracetic acid (versene)
EFC	Eagle's MEM supplemented with foetal calf serum
EGF	epidermal growth factor
EMS	ethyl methane sulphonate
ETC	Eagle's MEM supplemented with tryptosephosphate (10%) and calf serum (10%)
FGF	fibroblast growth factor
FITC	fluoroscein isothiocyanate
G0-phase	the resting stage of the cell cycle
G1-phase	the first gap in the cell cycle (between M and S)
G2-phase	the second gap in the cell cycle (between S and M)
GMP, GDP, GTP	guanosine monophosphate, diphosphate, triphosphate
GS-filter	0.22 μm membrane filter supplied by Millipore Corp.
HAT medium	medium supplemented with hypoxanthine, aminopterin and thymidine
HAU	haemagglutining unit
Hep cells	human epithelial cells
HPRT	hypoxanthine phosphoribosyl transferase
HSV	herpex simplex virus
HTC cells	hepatoma tissue culture cells
M	mitosis
MEM	minimum essential medium
Methocel	carboxymethyl cellulose type MC (4000 centripoises) sold by Dow Chemical Co.
MNNG	N-methyl-N-nitro-N-nitrosoguanidine
m.o.i.	multiplicity of infection
MOPC	mineral oil induced plasmacytoma
MSE	Medical and Scientific Equipment Ltd.
MVM	minute virus of mouse
NCTC	National Cancer Tissue Culture
NDV	Newcastle disease virus
PBS	phosphate buffered saline

PCA	perchloric acid
p.f.u.	plaque forming unit
PHA	phytohaemagglutinin
PPLO	pleuropneumonia-like organism
PPO	diphenyloxazole
p.s.i.	pounds per square inch (14.7 p.s.i. = 1 atmosphere = 10^5 Pascal)
PyY cells	polyoma transformed BHK cells
RPMI	Roswell Park Memorial Institute
RSV	Rous sarcoma virus
SDS	sodium dodecyl sulphate
S-phase	DNA synthetic phase of cell cycle
SSC	saline sodium citrate
SV40	simian virus 40
T-antigen	transplantation antigen
TC	tissue culture
TCA	trichloracetic acid
tG1, tG2, tS and tM	the time for the cell phases G1, G2, S and M, resp.
TK	thymidine kinase
TRITC	tetramethylrhodamine isothiocyanate
ts	temperature sensitive
UMP, UDP, UTP	uridine monophosphate, diphosphate, triphosphate
VSV	vesicular stomatitis virus
7X	a detergent

References

AARONSON, S.A. and G.J. TODARO (1968) Science *162*, 1024.

ABERCROMBIE, M. and J.E.M. HEAYSMAN (1954) Exp. Cell Res. *6*, 293.

ABERCROMBIE, M. and E.J. AMBROSE (1962) Cancer Res. *22*, 525.

ABO-DARUB, J.M. (1977) Ph.D. Thesis, University of Glasgow.

ABO-DARUB, J.M., R. MACKIE and J.D. PITTS (1978) Bull. Cancer *63*, 357.

ADAMS, R.L.P., R. ABRAMS and I. LIEBERMAN (1966) J. Biol. Chem. *241*, 903.

ADAMS, R.L.P. and J.G. LINDSAY (1967) J. Biol. Chem. *242*, 1314.

ADAMS, R.L.P. (1968) FEBS Lett. *2*, 91.

ADAMS, R.L.P. (1969a) Exp. Cell Res. *56*, 49.

ADAMS, R.L.P. (1969b) Exp. Cell Res. *56*, 55.

ADAMS, R.L.P., S. BERRYMAN and R. THOMSON (1971) Biochim. Biophys. Acta *240*, 455.

AHERNE, W.A., R.S. CAMPLEJOHN and N.A. WRIGHT (1977) Cell Population Kinetics (E. Arnold).

ALEXANDER, S.S., G. COLONNA, K.M. YAMADA, I. PASTAN and H. EDELHOCH (1978) J. Biol. Chem. *253*, 5820.

ALT, F.W., R.E. KELLEMS and R.T. SCHIMKE (1976) J. Biol. Chem. *251*, 3063.

ALT, F.W., R.E. KELLEMS, J.R. BERTINO and R.T. SCHIMKE (1978) J. Biol. Chem. *253*, 1357.

ANTONIADES, H.N. and C.D. SCHER (1977) Proc. Nat. Acad. Sci. U.S.A. *74*, 1973.

APOSHIAN, H.V. and A. KORNBERG (1962) J. Biol. Chem. *237*, 519.

ARMSTRONG, D. (1966) Proc. Soc. Exp. Biol. Med. *122*, 475.

ARMSTRONG, D. (1973) In: Contamination in Tissue Culture, Fogh, J. (ed.) (Academic Press, New York) p. 51.

ARRIGHI, F.E. and T.C. HSU (1965) Exp. Cell Res. *39*, 305.

AUB, J.C., C. TIESLAU and A. LANKESTER (1963) Proc. Nat. Acad. Sci. U.S.A. *50*, 613.

AUGUSTI-TOCCO, G. and G. SATO (1969) Proc. Nat. Acad. Sci. U.S.A. *64*, 311.

BAKER, J.B., R.L. SIMMER, K.C. GLENN and D.D. CUNNINGHAM (1979) Nature *278*, 743.

BALLARD, P.L. and G.M. TOMKINS (1969) Nature *224*, 344.

BALTIMORE, D. and D. SMOLER (1971) Proc. Nat. Acad. Sci. U.S.A. *68*, 1507.

271

BARNSTABLE, C.J., W.F. BODMER, G. BROWN, G. GALFRE, C. MILSTEIN, A.F. WILLIAMS and A. ZIEGLER (1978) Cell *14*, 9.

BARON, S. and A. ISAACS (1962) Brit. Med. J. *1*, 18.

BARON, S. (1969) *In:* Fundamental Techniques in Virology, Habel, K. and Satzman, N.P. (eds.) (Academic Press, New York) p. 399.

BASERGA, R. (1961) J. Hist. Cyt. *9*, 586.

BASERGA, R. and D. MALAMUD (1969) Autoradiography: Techniques and Applications (Harper, New York) p. 54.

BASERGA, R. (1978) J. Cell Physiol. *95*, 377.

BASILICO, C. (1977) Adv. Cancer Res., Klein, G. and Weinhous, S. (eds.) (Academic Press, New York) Vol. 24, p. 223.

BASILICO, C. (1978) J. Cell Physiol. *95*, 367.

BASILICO, C. and D. ZOUZIAS (1976) Proc. Nat. Acad. Sci. U.S.A. *73*, 1931.

BAUMUNK, C.N. and D.L. FRIEDMAN (1971) Cancer Res. *31*, 1930.

BELLO, L.J. (1974) Exp. Cell Res. *89*, 263.

BENDA, P., J. LIGHTBODY, G.H. SATO, L. LEVINE and H. SWEET (1968) Science *161*, 370.

BENNETT, M.V.L. (1973) Fed. Proc. *32*, 65.

BERNSTINE, E.G., M.L. HOOPER, S. GRANDCHAMP and B. EPHRUSSI (1973) Proc. Nat. Acad. Sci. U.S.A. *70*, 3899.

BISWAS, D.K., J. LYONS and A.H. TASHJIAN (1977) Cell *11*, 431.

BJURSELL, G. and P. REICHARD (1973) J. Biol. Chem. *218*, 3904.

BLACK, P.H., E.M. CRAWFORD and L.U. CRAWFORD (1964) Virology *24*, 381.

BLAKLEY, R.L. (1969) *In:* The Biochemistry of Folic Acid and Related Pteridines, Frontiers of Biology, Vol. 13, Neuberger, A. and Tatum, E.L. (eds.) (North Holland Publ. Co., Amsterdam).

BOLEN, J.B. and G.L. SMITH (1977) J. Cell Physiol. *91*, 441.

BOONE, C.W., G.S. HARELL and H.E. BOND (1968) J. Cell Biol. *36*, 369.

BOOTSMA, D., L. BUDKE and O. VOS (1964) Exp. Cell Res. *33*, 301.

BORSA, J. and C.F. WHITMORE (1969) Mol. Pharmacol. *5*, 318.

BOSTOCK, C.J., D.M. PRESCOTT and J.B. KIRKPATRICK (1971) Exp. Cell Res. *68*, 163.

BOURGEOIS, S. and R.F. NEWBY (1977) Cell *11*, 423.

BOYUM, A. (1968) Scand. J. Clin. Invest. *21*, (Suppl. 97), 31.

BRADBURY, S. (1976) The Optical Microscope in Biology (Edward Arnold, London).

BRENT, J.P., J.A.V. BUTLER and A.R. CRATHORN (1965) Nature *207*, 176.

BRENT, J.P. (1971) Cell Tissue Kinet. *4*, 297.

BROOKS, R.F. (1975) J. Cell Physiol. *86*, 369.

BROOKS, R.F. (1976) Nature *260*, 248.

BROWN, N.C. and P. REICHARD (1969) J. Mol. Biol. *46*, 39.

BUCKINGHAM, M.E. (1977) *In:* Biochemistry of Cell Differentiation, Vol. II, Paul, J. (ed.) (University Park Press, Baltimore) p. 269.

BUONASSISI, V., G. SATO and A.I. COHEN (1962) Proc. Nat. Acad. Sci. U.S.A. *48*, 1184.

BURGER, M.M. (1970) Nature *227*, 170.

BURK, R.R. (1970) Exp. Cell Res. 63, 309.

BURK, R.R. (1976) Exp. Cell Res. 101, 293.

BURSKIRK, H.H. (1967) Appl. Microbiol. 15, 1442.

BURTON, K. (1956) Biochem. J., 62, 315.

CAIRNS, J. (1966) J. Mol. Biol. 15, 372.

CAIRNS, J. (1972) Harvey Lects. 66, 1.

CALISSANO, P., G. MONACO, L. CASTELLANI, D. MERCANTI and A. LEVI (1978) Proc. Nat. Acad. Sci. U.S.A. 75, 2210.

CALLAN, H.G. (1972) Proc. Roy. Soc. Lond. B. 181, 19.

CARO, L. (1966) Progr. Biophys. 16, 171.

CARPENTER, G., L. KING and S. COHEN (1978) Nature 276, 409.

CASKEY, C.T. and G.D. KRUH (1979) Cell 16, 1.

CASPERSSON, T., L. ZECH, C. JOHNSSON and E.J. MODEST (1970a) Chromosoma 30, 215.

CASPERSSON, T., M. HUTHEN, J. LINDSTEN and L. ZECH (1970b) Exp. Cell Res. 63, 240.

CASPERSSON, T., K.G. LOMAK and L. ZECH (1971) Hereditas 67, 89.

CLAUSEN, J. (1969) Laboratory Techniques in Biochemistry and Molecular Biology, Work, T.S. and Work, E. (eds.) (North-Holland Publ. Co., Amsterdam) p. 423.

CLEAVER, J.E. (1965) Exp. Cell Res. 39, 697.

CLEAVER, J.E. and R.M. HOLFORD (1965) Biochim. Biophys. Acta 103, 654.

CLEAVER, J.E. (1967) Thymidine Metabolism and Cell Kinetics (North Holland Publ. Co., Amsterdam).

CLEAVER, J.E. (1969a) Nature 218, 652.

CLEAVER, J.E. (1969b) Radiat. Res. 37, 334.

COMINGS, D.E., E. AVELINO, T.A. OKADA and H.E. WYANDT (1973) Exp. Cell Res. 77, 469.

COOPER, J.E.K. (1973) In: Tissue Culture: Methods and Applications, Kruse, P.F. and Patterson, M.K. (eds.) (Academic Press, New York) p. 266.

COOPER, R.A., S. PERRY and T.R. BREITMAN (1966) Cancer Res. 20, 2265.

CORY, J.G., M.M. MANSELL and T.W. WHITFORD (1976) Adv. Enz. Regulat. 14, 45.

CREAGAN, R.P. and F.H. RUDDLE (1977) In: Molecular Structure of Human Chromosomes, Yunis, J.J. (ed.) (Academic Press, New York) p. 89.

CREASEY, W.A. and M.C. MARKIW (1965) Biochim. Biophys. Acta 103, 635.

CRESS, A.E. and E.W. GERNER (1977) Exp. Cell Res. 110, 347.

CROCE, C.M., W. SAWICKI, D. KRITCHEVSKY and H. KOPROWSKI (1971) Exp. Cell Res. 67, 427.

CRUICKSHANK, C.N.D., J.R. COOPER and M.B. CONRAN (1959) Exp. Cell Res. 16, 695.

DAIN, A.R. and J.G. HALL (1967) Vox Sang. 13, 281.

DALTON, C.C. and H.E. STREET (1976) In vitro 12, 485.

DAS, M. and C.F. FOX (1978) Proc. Nat. Acad. Sci. U.S.A. 75, 2644.

DAVIDSON, R.L., K.A. O'MALLEY and T.B. WHEELER (1976) Somatic Cell Genet. 2, 271.

DAWSON, K.B., E.O. FIELD and G.W. STEVENS (1962) Nature 195, 210.

DEAN, P. and J. JETT (1974) J. Cell Biol. *60*, 523.

DEFENDI, V. and L.A. MANSON (1963) Nature *198*, 359.

DEIBEL, R. and J. HOTCHIN (1961) Virology *14*, 66.

DOLFINI, S. (1971) *In:* Invertebrate Tissue Culture, Vol. 1 Vago, C. (ed.) (Academic Press, New York) p. 247.

DOLFINI, S., A.M. COURGEON and L. TIEPOLO (1970) Experientia *26*, 1020.

DONIACH, I. and S.R. PELC (1950) Br. J. Radiol. *23*, 184.

DULBECCO, R. (1969) Science *166*, 962.

DULBECCO, R. (1970) Nature *227*, 802.

DULBECCO, R. and G. FREEMAN (1969) Virology *8*, 396.

DULBECCO, R. and J. ELKINGTON (1973) Nature *246*, 197.

EAGLE, H. (1955a) Science, *122*, 501.

EAGLE, H. (1955b) J. Biol. Chem. *214*, 839.

EAGLE, H. (1959) Science *130*, 432.

EAGLE, H. (1971) Science *174*, 500.

EAGLE, H., K.A. PIEZ and V.I. OYAMA (1961) J. Biol. Chem. *236*, 1425.

EAGLE, H. and K.A. PIEZ (1962) J. Exp. Med. *116*, 29.

EARLE, W.R. (1943) J. Nat. Cancer Inst. *3*, 555.

EARLE, W.R., J.C. BRYANT and E.L. SHILLING (1954) Ann. N.Y. Acad. Sci. *58*, 1000.

ECHALIER, G. and A. OHANESSIAN (1969) Compt. Rend. Acad. Sci. *268*, 1771.

ENDERS, J.F., T.H. WELLER and F.C. ROBBINS (1949) Science *109*, 85.

ENGER, M.D., R.A. TOBEY and A.G. SAPONARA (1968) J. Cell Biol. *36*, 583.

ENGER, M.D. and R.A. TOBEY (1972) Biochemistry *11*, 269.

ENGLEBERG, J. (1961) Exp. Cell Res. *23*, 218.

ESKO, J.D. and C.R.H. RAETZ (1978) Proc. Nat Acad. Sci. U.S.A. *75*, 1190.

EVANS, M.J. (1972) J. Embryol. Exp. Morphol. *28*, 163.

EVANS, V.J., J.C. BRYANT, H.A. KERR and E.L. SHILLING (1964) Exp. Cell Res. *36*, 439.

EVERHART, L.F. (1972) Exp. Cell Res. *74*, 311.

FENNER, F., B.R. McAUSLAND, C.A. MIMS, J. SAMBROOK and D.O. WHITE (1974) The Biology of Animal Viruses (Academic Press Inc. New York)

FINBOW, M.E. and J.D. PITTS (1979) Manuscript in preparation.

FIRKET, H. (1965) *In:* Cells and Tissues in Culture, Vol. 1, Willmer, E.V. (ed.) (Academic Press Inc. New York), p. 201.

FIRKET, H. and P. MAHIEU (1966) Exp. Cell Res. *45*, 11.

FISHER, H.W., T.T. PUCK and G. SATO (1958) Proc. Nat. Acad. Sci. U.S.A. *44*, 4.

FISZMAN, M.Y. and P. FUCHS (1975) Nature *254*, 429.

FITZGERALD, P.J., M.G. ORD and L.A. STOCKEN (1961) Nature *189*, 55.

FOGH, J. and C. HACKER (1960) Exp. Cell Res. *21*, 242.

FOGH, J. and H. FOGH (1968) Proc. Soc. Exp. Biol. Med. *117*, 899.

FOGH, J. (1973) Contamination in Tissue Culture (Academic Press Inc., New York).

FOLKMAN, J. and A. MOSCONA (1978) Nature *273*, 345.

FORSDYKE, D.R. (1968) Biochem. J. *107*, 197.

FRAENKEL-CONRAT, H. (1974) Comprehensive Virology *1*, Fraenkel-Conrat, H. and Wagner, R.R. (eds.) (Elsevier Publ., Amsterdam) p. 3.

FRESHNEY, R.I. (1975) *In:* Laboratory Manual of Cell Biology, Hall, D. and Hawkins, S. (eds) (English Universities Press, London) p. 245.

FRESHNEY, R.I. (1979) *In:* Brain Tumours: Scientific Basis, Clinical Investigation and Current Therapy, Thomas, D.G.T. and Graham. D.I. (eds.) (Butterworths).

FRIDLAND, R. (1974) Cancer Res. *34*, 1883.

FRIDLAND, A. and J.P. BRENT (1975) Eur. J. Biochem. *57*, 379.

FRIEND, C., M.C. PATULEIA and E. DE HARVEN (1966) Nat. Cancer Inst. Monogr. *22*, 505.

FRIEND, C., W. SCHER, J.G. HOLLAND and T. SATO (1971) Proc. Nat. Acad. Sci. U.S.A. *68*, 378.

FULWYLER, M.J. (1965) Science *150*, 910.

FURLONG, N.B., W.B. NOVAK and J.E. STUBBLEFIELD (1973) Cell Tissue Kinet. *6*, 303.

GAMBARI, R., M. TERADA, A. BANK, R.A. RIFKIND and P.A. MARKS (1978) Proc. Nat. Acad. Sci. U.S.A. *75*, 3801.

GARDNER, R.S. (1969) J. Cell Biol. *42*, 320.

GELBARD, A.S., J.H. KIM and A.G. PEREZ (1969) Biochim. Biophys. Acta *182*, 564.

GELFANT, S. (1959) Exp. Cell Res. *16*, 527.

GELFANT, S. (1963) Intern. Rev. Cytol. *14*, 1.

GENTRY, G.A., P.A. MORSE, D.H. IVES, R. GEBERT and R. VAN POTTER (1965) Cancer Res. *25*, 509.

GEY, G.O. (1955) Harvey Lects. *50*, 154.

GILBERT, C.W., S. MULDAL and L.G. LAJTHA (1965) Nature *208*, 159.

GILES, R.E. and F.H. RUDDLE (1973) *In:* Tissue Culture: Methods and Applications, Kruse, P.F. and Patterson, M.K. (eds.) (Academic Press, New York), p. 475.

GILL, G. et al. (1978) Sixth Cold Spring Harbor Meeting on Cell Proliferation, in press.

GIRARD, M., S. MONTEUIL, M. FISZMAN, M. MARX and G. DANGLOT (1975) *In:* Tumour Virus – Host Cell Interaction, Kolber, A. (ed.) (Plenum Press, New York) p. 69.

GLENN, K.C. and D.D. CUNNINGHAM (1979) Nature *278*, 711.

GODBER, G. (1975) Report of the Working Party on the Laboratory Use of Dangerous Pathogens. H.M. Stationery Office, Cmnd. 6054.

GOSS, S.J. and H. HARRIS (1975) Nature *255*, 680.

GRACE, T.D.C. (1962) Nature *195*, 788.

GRACE, T.C.D. (1966) Nature *211*, 366.

GRAY, P. (1972) The use of the Microscope (McGraw Hill, New York).

GREEN, M. (1970) Ann. Rev. Biochem. *39*, 701.

GREEN, H. and M. MEUTH (1974) Cell *3*, 127.

GREEN, H. (1977) Cell *11*, 405.

GREEN, H. (1978) Cell *15*, 801.

GRESSER, I. (1961) Proc. Soc. Exp. Biol. Med. *108*, 799.

GUNER, M., R.I. FRESHNEY, D. MORGAN, M.G. FRESHNEY, D.G.T. THOMAS and D.I. GRAHAM (1977) Br. J. Cancer *35*, 439.

GWATKIN, R.B.L. (1973) *In:* Tissue Culture: Methods and Applications, Kruse, P.F. and Patterson, M.K. (eds) (Academic Press, New York), p. 3.

HABEL, K. and N.P. SALZMAN (1969) Fundamental Techniques in Virology (Academic Press Inc., New York).

HAIGLER, H., J.F. ASH, S.J. SINGER and S. COHEN (1978) Proc. Nat. Acad. Sci. U.S.A. *75*, 3317.

HAKALA, M.T. and T. ISHIHARA (1962) Cancer Res. *22*, 987.

HAKALA, M.T., J.F. HOLLAND and H.S. HOROSZEWICZ (1963) Biochem. Biophys. Res. Comm. *11*, 466.

HALL, D. and S. HAWKINS, Laboratory Manual of Cell Biology (The English Universities Press Ltd., London).

HAM, R.G. (1963) Exp. Cell Res. *29*, 515.

HAM, R.G. (1965) Proc. Nat. Acad. Sci. U.S.A. *53*, 288.

HAM, R.G. and T.T. PUCK (1967) Methods in Enzymology, Vol. V, Colowick, S.P. and Kaplan, N.O. (eds.) (Academic Press, Inc., New York), p. 90.

HAM, R.G. and W.L. MCKEEHAN (1978) In vitro *14*, 11.

HAND, R., W.D. ENSMINGER and I. TAMM (1971) Virology *44*, 527.

HAND, R. and I. TAMM (1974) *In:* Cell Cycle Controls, Padilla, G.M., Cameron, I.L. and Zimmerman, A. (eds.) (Academic Press, New York), p. 273.

HARADA, F., R.C. SAWYER and J.E. DAHLBERG (1975) J. Biol. Chem. *250*, 3487.

HARRIS, H. and J.F. WATKINS (1965) Nature *205*, 640.

HARRIS, H. (1970) Cell Fusion (Clarendon Press, Oxford).

HARRIS, M. (1964) Cell Culture and Somatic Variation (Holt, Rinehart and Winston, Inc., New York).

HARRISON, P.R. (1976) Nature *262*, 353.

HARRISON, P.R. (1977) *In:* Biochemistry of Cell Differentiation, Vol. II, Paul, J. (ed.) (University Park Press, Baltimore), p. 227.

HARRISON, P.R., T. RUTHERFORD, D. CONKIE, N. AFFARA, J. SOMMERVILLE, P. HISSEY and J. PAUL (1978) Cell *14*, 61.

HATCHAND, C.G. and C.A. PARKER (1956) Proc. Roy. Soc. London (A) *235*, 518.

HAYASHI, I. and G.H. SATO (1976) Nature *259*, 132.

HAYASHI, I., J. LARNER and G. SATO (1978) In vitro *14*, 23.

HAYFLICK, L. and P.S. MOORHEAD (1961) Exp. Cell Res. *25*, 585.

HAYFLICK, L. (1965a) Exp. Cell Res. *37*, 614.

HAYFLICK, L. (1965b) Tex. Rep. Biol. Med. *23*, Suppl. 1. 285.

HEALY, G.M. and R.C. PARKER (1966a) J. Cell Biol. *30*, 531.

HEALY, G.M. and R.C. PARKER (1966b) J. Cell Biol. *30*, 539.

HELLER, R. (1953) Ann. Sci. Nat. (Bot. Biol. Veg.) Paris *14*, 1.

HELLMAN, A., M.N. ORMAN and R. POLLACK (1973) Biohazards in Biological Research (Cold Spring Harbor Labs.).

HELLSTRÖM, I. (1967) Int. J. Cancer 2, 65.

HELLSTRÖM, I. and K.E. HELLSTRÖM (1971) In vitro Methods in Cell Mediated Immunity, Bloom, B.B. and Glade, P.R. (eds.) (Academic Press, New York), p. 409.

HELLSTRÖM, U., S. HAMMARSTRÖM and M.-L. DILLNER (1976) Scand. J. Immunol. 5, Suppl. 5, 45.

HENLE, G. and W. HENLE (1969) J. Bacteriol. 91, 1248.

HERZENBERG, L.A., L.G. SWEET and L.A. HERZENBERG (1976) Scient. Am. 234 (3), 108.

HILFER, S.R. (1973) In: Tissue Culture, Methods and Applications, Kruse, P.F. and Patterson, M.K. (eds.) (Academic Press Inc., New York) p. 16.

HINEGARDNER, R.T. (1971) Anal. Biochem. 39, 197.

HIRSCHBERG, H., H. SKARE and E. THORSBY (1977) J. Immunol. Meth. 16, 131.

HIRSCHMAN, S.Z. and G. FELSENFELD (1966) J. Mol. Biol. 16, 347.

HIRT, B. (1967) J. Mol. Biol. 26, 365.

HOGAN, B.L.M. (1977) In: Biochemistry of Cell Differentiation, Vol. II, Paul, J. (ed.) (University Park Press, Baltimore), p. 333.

HOLLEY, R.W. and J.A. KIERNAN (1968) Proc. Nat. Acad. Sci. U.S.A. 60, 300.

HOLLEY, R.W. (1975) Nature 258, 487.

HOLTZER, H., J. BIEHL, G. YFOH, R. MEGANATHA and A. KAJI (1975) Proc. Nat. Acad. Sci. U.S.A. 72, 4051.

HORIBATA, K. and A.W. HARRIS (1970) Exp. Cell Res. 60, 61.

HORIKAWA, M., and A.S. FOX (1964) Science 145, 1437.

HOSAKA, Y. and Y. KOSHI (1968) Virology 34, 419.

HOUSE, W. and A. WADDELL (1967) J. Pathol. Bacteriol. 93, 125.

HOUSE, W. (1973) In: Tissue Culture, Methods and Applications, Kruse, P.F. and Patterson, M.K. (eds.) (Academic Press Inc., New York), p. 338.

HOWARD, A. and S.R. PELC (1953) Heredity (Suppl). 6, 261.

HOZUMI, V. and S. TONEGAWA (1976) Proc. Nat. Acad. Sci. U.S.A. 73, 3628.

HSU, T.C. and D.S. KELLOGG (1960) J. Nat. Canc. Inst. 25, 221.

HSU, T.C. (1973) In: Tissue Culture, Methods and Applications, Kruse, P.F. and Patterson, M.K. (eds.) (Academic Press Inc., New York), p. 764.

HUBERMAN, J.A. and A.D. RIGGS (1968) J. Mol. Biol. 32, 327.

HUDSON, S., W.B. UPHOLT, J. DEVINNEY and J. VINOGRAD (1969) Proc. Nat. Acad. Sci. U.S.A. 62, 813.

HUNTER, D. and R.R. BOMFORD (1968) Hutchison's Clinical Methods (Bailliere, Tindall and Cassell, Ltd., London), p. 124.

HYNES, R.O. and J.M. BYE (1974) Cell 3, 113.

ILLMENSEE, K. and L.C. STEVENS (1979) Scient. Am. 240, 87.

IVES, D.H., P.A. MORSE and R. VAN POTTER (1963) J. Biol. Chem. 238, 1467.

IWAKATA, S. and J.T. GRACE (1964) N.Y. J. Med. 18, 2279.

IYER, V.N. and W. SZYBALSKI (1964) Science 145, 55.

JOHNSON, R.W. and M.D. ORLANDO (1967) Appl. Microbiol. *15*, 209.

KAJIWARA, K. and G.C. MUELLER (1964) Biochim. Biophys. Acta *91*, 486.

KAO, F.T. and T.T. PUCK (1968) Proc. Nat. Acad. Sci. U.S.A. *60*, 1275.

KAPLAN, A.S. (1969) *In:* Fundamental Techniques in Virology, Habel, K. and Salzman, N.P. (eds.) (Academic Press, New York), p. 487.

KEAY, L., S.A. WEISS, N. CIRULIS and B.S. WILDI (1972) In vitro *8*, 19.

KEIR, H.M., J. HAY, J.M. MORRISON and J. SUBAK-SHARPE (1966) Nature *210*, 369.

KENNY, G.E. and M.E. POLLOCK (1963) J. Infect. Dis. *112*, 7.

KENNY, G.E. (1973) *In:* Contamination in Tissue Culture, Fogh, J. (ed.) (Academic Press, New York), p. 107.

KERR, I.M., R.E. BROWN and L.A. BALL (1974) Nature *250*, 57.

KISHIMOTO, S. and I. LIEBERMAN (1964) Exp. Cell Res. *36*, 92.

KISSANE, J.M. and E. ROBINS (1958) J. Biol, Chem. *233*, 184.

KLEBE, R.J. (1974) Nature *250*, 248.

KLEVECZ, R.R. (1969) Science *166*, 1536.

KLEVECZ, R.R., L.N. KAPP and J.A. REMINGTON (1974) *In:* Control of Proliferation of Animal Cells, Clarkson, B. and Baserga, R. (eds.) (Cold Spring Harbor Press), p. 817.

KLEVECZ, R.R. (1975) Methods in Cell Biol., *10*, Prescott, D.M. (ed.) (Academic Press Inc., New York), p. 157.

KLEVECZ, R.R., B.A. KENISTON and L.L. DEAVEN (1975) Cell *5*, 195.

KLOTZ, L.C. and B.H. ZIMM (1972) J. Mol. Biol. *72*, 779.

KNAZEK, R.A. and P.M. GULLINO (1973) *In:* Tissue Culture, Methods and Applications, Kruse, P.F. and Patterson, M.K. (eds.) (Academic Press Inc., New York), p. 321.

KNAZEK, R.A., P.O. KOHLER and P.M. GULLINO (1974) Exp. Cell Res. *84*, 251.

KNIAZEFF, A.J. (1973) *In:* Contamination in Tissue Culture, Fogh, J. (ed.) (Academic Press, New York), p. 233.

KOHLER, G. and C. MILSTEIN (1975) Nature *256*, 495.

KOHLER, G. and C. MILSTEIN (1976) Eur J. Immunol. *6*, 511.

KOWALSKI, J. and W.P. CHEEVERS (1976) J. Mol. Biol. *104*, 603.

KRAEMER, P.M., L.L. DEAVEN, H.A. CRISSMAN, J.A. STEINKAMP and D.F. PETERSON (1973) Cold Spring Harbor Symp. Quant. Biol. *38*, 133.

KRAKOFF, I.H., N.C. BROWN and P. REICHARD (1968) Cancer Res. *28*, 1559.

KUBITSCHEK, H.E. (1966) Nature *209*, 1039.

KUCHLER, R.J. (1977) Biochemical Methods in Cell Culture and Virology (Dowdon, Hutchinson and Ross, Stroudsburgh, Pa.).

KUMMER, D., F. KAML, W. HEITLAND and E. JACOB (1978) Z. Krebsforsch. *91*, 23.

LARK, K.G., R. CONSIGLI and A. TOLIVAR (1971) J. Mol. Biol. *58*, 873.

LAVAPPA, K.S. (1978) In vitro, *14*, 469.

LAZARIDES, E. and J.P. REVEL (1979) Scient. Am. *240* (5), 88.

LEAKE, R.E. (1978) Personal communication.

LEE, M.J., M.H. VAUGHAN and R. ABRAMS (1970) J. Biol. Chem. *245*, 4525.

LEFFERT, H.L., T. MORAN, R. BOORSTEIN and K.S. KOCH (1977) Nature *767*, 58.

LENNARTZ, K.J. and W. MAURER (1964) Z. Zellforsch. *63*, 478.

LETT, J.T. and C. SUN (1970) Radiat. Res. *44*, 771.

LEY, K.D. and R.A. TOBEY (1970) J. Cell Biol. *47*, 453.

LIEBERMAN, I. and P. OVE (1957) Biochim. Biophys. Acta *25*, 449.

LIEBERMAN, I. and P. OVE (1962) J. Biol. Chem. *237*, 1634.

LIEBERMAN, I., R. ABRAMS, N. HUNT and P. OVE (1963) J. Biol. Chem. *238*, 3955.

LIGHTBODY, J., S.E. PFEIFFER, P.L. KORNBLITT and H.R. HERSCHMAN (1970) J. Neurobiol. *1*, 411.

LIN, C.C., I.A. UCHIDA and E. BYRNES (1971) Can. J. Genet. Cytol. *13*, 361.

LINDSAY, J.G. (1969) Ph.D. Thesis, Glasgow University.

LINDSAY, J.G., S. BERRYMAN and R.L.P. ADAMS (1970) Biochem. J. *119*, 839.

LINSLEY, P.S., C. BLIFELD, M. WRANN and C.F. FOX (1979) Nature *278*, 745.

LIPKIN, M. (1971) *In:* Cell Cycle and Cancer, Baserga, R. (ed.) (Marcel Dekker Inc., New York), p. 1.

LIPPMAN, M., M.E. MONACO and G. BOLAN (1977) Cancer Res. *37*, 1901.

LISKAY, R.M. and D.M. PRESCOTT (1978) Proc. Nat. Acad. Sci. U.S.A. *75*, 2873.

LITTLEFIELD, J.W. (1964) Science *145*, 709.

LITTLEFIELD, J.W. (1966a) Biochim. Biophys. Acta *114*, 398.

LITTLEFIELD, J.W. (1966b) Exp. Cell Res. *41*, 190.

LITTLEFIELD, J.W. and C. BASILICO (1966) Nature *211*, 250.

LITTLEFIELD, J.W. (1969) Proc. Nat. Acad. Sci. U.S.A. *62*, 88.

LITTLEFIELD, J.W. and S. GOLDSTEIN (1970) In vitro *6*, 21.

LOCKART, R.Z. and H. EAGLE (1959) Science *129*, 252.

LOOMIS, W.F., J.P. WAHRMANN and D. LUZZATI (1973) Proc. Nat. Acad. Sci. U.S.A. *70*, 425.

LOVELOCK, J.E. and M.W.H. BISHOP (1959) Nature *183*, 1394.

LOWRY, O.H., N.J. ROSEBROUGH, A.L. FARR and R.J. RANDALL (1951) J. Biol. Chem. *193*, 265.

MACDONALD, H.R. and R.G. MILLER (1970) Biophys. J. *10*, 834.

MACPHERSON, I. (1973a) *In:* Tissue Culture, Methods and Applications, Kruse, P.F. and Patterson, M.K. (eds.) (Academic Press Inc., New York) p. 241.

MACPHERSON, I. (1973b) Ibid, p. 276.

MACPHERSON, I.A. and M. STOKER (1962) Virology *16*, 147.

MACPHERSON, I.A. and L. MONTAGNIER (1964) Virology *23*, 291.

MACIEIRA-COELHO, A., J. PONTEN and L. PHILIPSON (1966) Exp. Cell Res. *43*, 20.

MARCHI, A. and K.S. RAI (1978) Can. J. Genet. Cytol. *20*, 243.

MARTIN, G.M. (1973) *In:* Tissue Culture, Methods and Applications, Kruse, P.F. and Patterson, M.K. (eds.) (Academic Press Inc., New York) p. 39.

MARKS, P.A. and R.A. RIFKIND (1978) Ann. Rev. Biochem. *47*, 419.

MARSHALL, R. (1972) Chromosoma *37*, 395.

MARTIN, G.R. and M.J. EVANS (1975a) Proc. Nat. Acad. Sci. U.S.A. 72, 1441.

MARTIN, G.R. and M.J. EVANS (1975b) Cell 6, 467.

MASSIE, H.R. (1972) In vitro 7, 191.

MATSUYA, Y. and H. GREEN (1969) Science 163, 697.

MATURO, J.M. and M.D. HOLLENBERG (1978) Proc. Nat. Acad. Sci. U.S.A. 75, 3070.

MAUEL, J. and V. DEFENDI (1971a) Exp. Cell Res. 65, 33.

MAUEL, J. and V. DEFENDI (1971b) J. Exp. Med. 134, 335.

MAYHEW, E. (1972) J. Cell Physiol. 79, 441.

MCBURNEY, M.W. and G.F. WHITMORE (1975) Cancer Res. 35, 586.

MCCOY, T.A., M. MAXWELL and P.F. KRUSE (1959) Proc. Soc. Exp. Biol. Med. 100, 115.

MELNICK, J.L. (1955) Ann. N.Y. Acad. Sci. 61, 754.

MERIGAN, T.C., D.F. GREGORY and J.K. PETRALLI (1966) Virology 29, 515.

MILLER, C.L. and F.H. RUDDLE (1978) Proc. Nat. Acad. Sci. U.S.A. 75, 3346.

MILLER, O.L., G.E. STONE and D.M. PRESCOTT (1964a) J. Cell Biol. 23, 654.

MILLER, O.L., G.E. STONE AND D.M. PRESCOTT (1964b) Methods in Cell Physiol., 1, Prescott, D.M. (ed.) (Academic Press Inc., New York) p. 371.

MILLER, R.G. and R.A. PHILLIPS (1969) J. Cell Physiol. 73, 191.

MILLER, Z., E. LOVELACE, M. GALLO and I. PASTAN (1975) Science 190, 1213.

MINOR, P.D. and J.A. SMITH (1974) Nature 248, 241.

MINTY, A.J., G.D. BIRNIE and J. PAUL (1978) Exp. Cell Res. 115, 1.

MINTZ, B. and K. ILLMENSEE (1975) Proc. Nat. Acad. Sci. U.S.A. 72, 3585.

MITCHISON, J.M. (1971) The Biology of the Cell Cycle (Cambridge University Press).

MITSUHASHI, J. and K. MARAMOROSCH (1964) Contrib. Boyce Thomson Inst. 22, 435.

MOENZ, W., A. VOKAER and R. KRAM (1975) Proc. Nat. Acad. Sci. U.S.A. 72, 1063.

MOORE, E.C. and R.B. HURLBERT (1966) J. Biol. Chem. 241, 4802.

MORGAN, J.M., H.J. MORTON and R.C. PARKER (1950) Proc. Soc. Exp. Biol. Med. 73, 1.

MORGAN, J.F., M.E. CAMPBELL and H.J. MORTON (1955) J. Nat. Cancer Inst. 16, 557.

MOORHEAD, P.S., P.C. NOWELL, W.J. MELLMAN, D.M. BATTIPS and D.A. HUNGERFORD (1960) Exp. Cell Res. 20, 613.

MORRIS, N.R. and G.A. FISCHER (1960) Biochim. Biophys. Acta 42, 183.

MORRIS, N.R., J.W. CRAMER and D. RENO (1967) Exp. Cell Res. 48, 216.

MORTON, H.J. (1970) In vitro 6, 89.

MUDD, S.H. and G.L. CANTONI (1964) In: Comprehensive Biochemistry, 15, Florkin, M. and Stotz, E.G. (eds.) (Elsevier, Amsterdam) p. 1.

MUELLER, G.C. and J. KAJIWARA (1966) Biochim. Biophys. Acta 119, 557.

MUELLER, G.C. (1971) In: The Cell Cycle and Cancer, Baserga, R. (ed.) (Marcel Dekker Inc., New York) p. 269.

MUKHERJEE, A.B., S. ORLOFF, J. DE BUTLER, T. TRICHE, P. LALLEY and J.B. SCHULMAN (1978) Proc. Nat. Acad. Sci. U.S.A. 75, 1361.

MURASHIGE, T. and F. SKOOG (1962) Physiol. Plant. 15, 473.

MURPHREE, S., E. STUBBLEFIELD and E.C. MOORE (1969) Exp. Cell Res. *58*, 118.

NADAI-GRINARD, B. (1978) Cell *15*, 855.

NAIRN, R.C. (ed.) (1969) Fluorescent Protein Tracing (Livingstone, Edinburgh) p. 61.

NEFF, J.M. and J.F. ENDERS (1968) Proc. Soc. Exp. Biol. Med. *127*, 260.

NELSON, D.S. (1976) Immunobiology of the Macrophage (Academic Press, New York).

NISHIMOTO, T., E. EILEN and C. BASILICO (1978) Cell *15*, 475.

NITSCH, J.P. and C. NITSCH (1956) Am. J. Bot. *43*, 839.

ODELL, W.D., P.L. RAYFORD and G.T. ROSS (1967) J. Lab. Clin. Med. *70*, 973.

OGBURN, C.A., K. BERG and K. PAUCKER (1973) J. Immunol. *111*, 1206.

OGLE, J.W., R.D. LANGE and C.D.R. DUNN (1978) In Vitro, *14*, 945.

OKADA, S. (1967) J. Cell Biol. *34*, 915.

OLDEN, K., L.H.E. HAHN and K.M. YAMADA (1979) Br. Soc. Cell Biol. Symp., in press.

OLDHAM, K.G. (1967) J. Labelled Comp. *IV*, 127.

ORKIN, S.H. (1978) In vitro *14*, 146.

PAPIOANNOU, V.E., M.W. MCBURNEY, R.L. GARDNER and M.J. EVANS (1975) Nature *258*, 70.

PARDEE, A.B. (1974) Proc. Nat. Acad. Sci. U.S.A. *71*, 1286.

PARKER, R.C. (1957) Special Publ. of N.Y. Acad. Sci. *5*, 303.

PARKER, R.C. (1959) Can. Cancer Conf. *3*, 189.

PATIL, S., S. MERRICK and H.A. LUTZ (1971) Science *173*, 821.

PATTERSON, M.K. and M.D. MAXWELL (1973) *In:* Tissue Culture, Methods and Applications, Kruse, P.F. and Patterson, M.K. (eds.) (Academic Press Inc., New York).

PATULEIA, M.C. and C. FRIEND (1967) Cancer Res. *27*, 726.

PAUL, J. (1975) Cell and Tissue Culture, 5th ed. (E. and S. Livingstone Ltd., Edinburgh).

PEARLSTEIN, E. (1976) Nature *262*, 497.

PEARSE, A.G.E. (1953) Histochemistry (Churchill, London).

PEGARARO, L. and G. BENZIO (1971) Experientia *27*, 33.

PERTOFT, H., T.C. LAURENT, T. LAAS and L. KAGEDAL (1978) Anal. Biochem. *88*, 271.

PETERSEN, D.F., E.C. ANDERSON and R.A. TOBEY (1968) *In:* Methods in Cell Physiology, Vol. 3, Prescott, D.M. (ed.) (Academic Press Inc., New York) p. 347.

PETERSON, D.M. and E.C. MOORE (1976) Biochim. Biophys. Acta *432*, 80.

PFEIFFER, S.E. and L.J. TOLMACH (1967) Nature *213*, 139.

PFEIFFER, S.E. and W. WECHSLER (1972) Proc. Nat. Acad. Sci. U.S.A. *69*, 2885.

PHARMACIA (1978) 'Cytodex™ 1' – Beaded Microcarriers for Cell Culture (Pharmacia Fine Chemicals A.B. Uppsala).

PITTS, J.D. (1971) Ciba Foundation Symposium on Growth Control in Cell Cultures, Wolstenholme, G.E.W., and Knight, J. (eds.) (Churchill Livingstone, London) p. 89.

PITTS, J.D. and M.E. FINBOW (1977) *In:* Intercellular Communication, De Mello, W.C. (ed.) (Plenum Press, New York) p. 61.

PITTS, J.D. and J.W. SIMMS (1977) Exp. Cell Res. *104*, 153.

POLLOCK, M.E. and G.E. KENNY (1963) Proc. Soc. Exp. Biol. Med. *112*, 176.

PONTECORVO, G., P.N. RIDDLE and A. HALES (1977) Nature *265*, 257.

PONTEN, J. (1973) *In:* Tissue Culture, Methods and Applications, Kruse, P.F. and Patterson, M.K. (eds.) (Academic Press Inc., New York) p. 50.

POTTER, M. (1972) Physiol. Rev. *52*, 631.

POTTER, M. (1975) *In:* Cancer – a Comprehensive Treatise I, Becker, F.F. (ed.) (Plenum Press, New York).

POTTER, M. (1976) *In:* Methods in Cancer Res. II, Busch, H. (ed.) (Academic Press, New York) p. 105.

PROP, F.J.A. and G.J. WIEPJES (1973) *In:* Tissue Culture, Methods and Applications, Kruse, P.F. and Patterson, M.K. (eds.) (Academic Press, Inc., New York) p. 21.

PUCK, T.T. (1964) Cold Spring Harbor Symp. Quant. Biol. *29*, 167.

PUCK, T.T., P.I. MARCUS and S.J. CIECIURA (1956) J. Exp. Med. *103*, 273.

PUCK, T.T. and J. STEFFEN (1963) Biophys. J. *3*, 379.

PUCK, T.T., P. SAUNDERS and D. PETERSEN (1964) Biophys. J. *4*, 441.

PUCK, T.T. (1972) The Mammalian Cell as a Microorganism (Holden Day Inc., San Francisco).

PURI, E.C. and D.C. TURNER (1978) Exp. Cell Res. *115*, 159.

RABBITTS, T.H. and C. MILSTEIN (1977) Cont. Top. Mol. Immunol. *6*, 117.

RABINOWITZ, Y. (1973) *In:* Tissue Culture, Methods and Applications, Kruse P.F. and Patterson, M.K. (eds.) (Academic Press Inc., New York) p. 25.

RABINOWITZ, Y. (1964) Blood *23*, 811.

RAFF, M.C., E. ABNEY, J.P. BROOKES and A. HORNBY-SMITH (1978) Cell *15*, 813.

REICHARD, P., Z.N. CANELLAKIS and E.S. CANELLAKIS (1961) J. Biol. Chem. *236*, 2514.

REINERT, J., Y.P.S. BAJAJ and B. ZBELL (1977) *In:* Plant Tissue and Cell Culture, 2nd ed. Street, H.E. (ed.) (Blackwell, Oxford) p. 389.

REVEL, M. and Y. GRONER (1978) Ann. Rev. Biochem. *47*, 1079.

RHEINWALD, J.G. and H. GREEN (1975) Cell *6*, 331.

RHEINWALD, J.G. and H. GREEN (1977) Nature *265*, 421.

RICCIUTI, F.C. and F.H. RUDDLE (1973) Nature New Biol. *241*, 180.

RICHAMAN, R.A., T.H. CLAUS, B.S.J. PILK and D.L. FRIEDMAN (1976) Proc. Nat. Acad. Sci. U.S.A. *73*, 3589.

RICHTER, C. and D. YAFFE (1970) Develop. Biol. *23*, 1.

RINGERTZ, N.R. and R.E. SAVAGE (1976) Cell Hybrids (Academic Press Inc., New York).

ROBBINS, E. and P.I. MARCUS (1964) Science *144*, 1152.

ROBBINS, E. and M.D. SCHARFF (1967) J. Cell Biol. *34*, 684.

RODWELL, A. (1969) *In:* The Mycoplasmatales and L-phase of Bacteria, Hayflick, L. (ed.) (Appleton, New York) p. 413.

RONALDONI, L.M.J. (1959) Exp. Cell Res. *16*, 477.

Roos, D. and J.A. Loos (1970) Biochim. Biophys Acta 222, 565.

Rosenthal, M.D., R.M. Wishnow and G.H. Sato (1970) J. Nat. Cancer Inst. 44, 1001.

Ross, R. and A. Vogel (1978) Cell 14, 203.

Rossini, M. and R. Baserga (1978) Biochemistry 17, 858.

Ruddle, F.H. (1973) Nature 242, 165.

Rudland, P.S. (1978) Nature 276, 113.

Rudland, P.S., D. Gospodarowitz and W. Siefert (1974) Nature 250, 741.

Ruekert, R.R. and G.C. Mueller (1960) Cancer Res. 20, 1584.

Russel, W.C., C. Newman and D.H. Williamson (1975) Nature 253, 461.

Rutherford, R.B. and R. Ross (1976) J. Cell Biol. 69, 196.

Rutter, W.J., R.L. Pictet and P.W. Morris (1973) Ann. Rev. Biochem. 42, 601.

Saborio, J.L., S.S. Pong and G. Koch (1974) J. Mol. Biol. 85, 195.

Sahyoun, N., R.A. Hock and M.D. Hollenberg (1978) Proc. Nat. Acad. Sci. U.S.A. 75, 1675.

Saito, H. and K.-I. Miura (1963) Biochim. Biophys. Acta 72, 619.

Samborn, R.C. and J.A. Haskell (1961) Proc. Int. Congr. Entomol. 83, 237.

Sanford, B.H., J.F. Codington, R.W. Jeanloz and P.D. Palmer (1973) J. Immunol. 110, 1233.

Sanford, K.K. (1973) In: Tissue Culture – Methods and Applications, Kruse P.F. and Patterson, M.K. (eds.) (Academic Press Inc., New York) p. 237.

Sanford, K.K., W.R. Earle and G.D. Likely (1948) J. Nat. Canc. Inst. 9, 229.

Sanford, K.K., W.R. Earle, V.J. Evans, H.K. Waltz and J.E. Shannon (1951) J. Nat. Cancer Inst. 11, 773.

Sanford, K.K., A.B. Covalesky, L.T. Dupree and W.R. Earle (1961) Exp. Cell Res. 23, 361.

Sato, G. (1973) Tissue Culture of the Nervous System (Plenum Press, New York).

Sauerborn, R., A. Balmain, K. Gaerttler and M. Stohr (1978) Cell Tissue Kinet. 11, 291.

Scher, W., D. Tsuei, S. Sassa, P. Price, N. Gabelman and C. Friend (1978) Proc. Nat. Acad. Sci. U.S.A. 75, 3851.

Schimke, R.T., R.J. Kaufman, J.H. Nunberg and S.L. Dana (1978) Cold Spring Harb. Symp. Quant. Biol. 43, 1297.

Schindler, R., N. Odastchenko, A. Grieder and L. Ramseier (1968) Exp. Cell. Res. 51, 1.

Schleicher, J.B. (1973) In: Tissue Culture: Methods and Applications, Kruse P.F. and Patterson, M.K. (eds.) (Academic Press Inc., New York) p. 333.

Schlessinger, J., Y. Schechter, M.C. Willingham and I. Pastan (1978) Proc. Nat. Acad. Sci. U.S.A. 75, 2659.

Schubert, D., A.J. Harris, S. Heinemann, Y. Kidokoro, J. Patrick and J.H. Steinbach (1973) Tissue Culture of the Nervous System, Sato, G. (ed.) (Plenum Press, New York) p. 55.

SEABRIGHT, M. (1971) Lancet *2*, 971.
SEEGMILLER, J.E., F.M. ROSENBLOOM and W.M. KELLEY (1967) Science *155*, 1682.
SCHECHTER, Y., L. HERNAEZ and P. CUATRECASAS (1978) Proc. Nat. Acad. Sci. U.S.A. *75*, 5788.
SHALL, S. and A.J. MCCLELLAND (1971) Nature New Biol. *279*, 59.
SHALL, S. (1973a) *In:* Tissue Culture – Methods and Applications, Kruse P.F. and Patterson M.K. (eds.) (Academic Press Inc., New York) p. 195.
SHALL, S. (1973b) Ibid, p. 198.
SHEININ, R. (1976) Cell *7*, 49.
SHIELDS, R., R.F. BROOKS, P.N. RIDDLE, D.F. CAPELLARO and D. DELIA (1978) Cell *15*, 469.
SIEGERS, M.P., J.C. SCHAER, H. HIRSIGER and R. SCHINDLER (1974) J. Cell Biol. *62*, 305.
SINCLAIR, R. and D.H.L. BISHOP (1965) Nature *205*, 1272.
SINCLAIR, W.K. (1965) Science *150*, 1729.
SINGH, K.R.P. (1967) Current Sci. *36*, 506.
SISKEN, J.E. and L. MORASCA (1965) J. Cell Biol. *25* (2, part 2), 179.
SKEA, B.R. and A.M. NEMETH (1969) Proc. Nat. Acad. Sci. U.S.A. *64*, 795.
SKOOG, K.L., B.A. NORDENSKJOLD and K.G. BJURSELL (1973) Eur. J. Biochem. *33*, 428.
SKOOG, K.L. and G. BJURSELL (1974) J. Biol. Chem. *249*, 6434.
SMITH, C.L. and P.P. DENDY (1962) Nature *193*, 555.
SMITH, J.A. and L. MARTIN (1973) Proc. Nat. Acad. Sci. U.S.A. *70*, 1263.
SMITH, J.A. and L. MARTIN (1974) *In:* Cell Cycle Controls, Padilla, G.M., Cameron, I.L. and Zimmerman, A. (eds.) (Academic Press Inc., New York) p. 43.
SMITH, H S., C.D. SCHER and G.J. TODARO (1970) Bacteriol. Proc. Abstr. *217*, 187.
SMITH, P.F. (1971) The Biology of Mycoplasmas (Academic Press, New York).
SOMERS, D.G., M.L. PEARSON and C.J. INGLES (1975) J. Biol. Chem. *250*, 4825.
SPENDLOVE, R.S., R.B. CROSBIE, S.F. HAYES and R.F. KEELER (1971) Proc. Soc. Exp. Biol. Med. *137*, 258.
STADLER, J.K. and E.A. ADELBERG (1972) Proc. Nat. Acad. Sci. U.S.A. *69*, 1929.
STANBRIDGE, E.J., L. HAYFLICK and F.T. PERKINS (1971) Nature New Biol. *232*, 242.
STEFFENSEN, D.M. (1977) *In:* Molecular Structure of Human Chromosomes, Yunis, J.J. (ed.) (Academic Press, New York) p. 59.
STERNBERGER, L.A., P.H. HARDY, J.J. CUCULIS and H.G. MEYER (1970) J. Histochem. Cytochem. *18*, 315.
STIMAC, E., D. HOUSEMAN and J.A. HUBERMAN (1977) J. Mol. Biol. *115*, 485.
STOCK, D.A. and G.A. GENTRY (1971) J. Gen. Microbiol. *65*, 105.
STOKER, M.G.P. and I.A. MACPHERSON (1961) Virology *14*, 359.
STOKER, M.G.P. and I.A. MACPHERSON (1964) Nature *203*, 1355.
STOKER, M.G.P. (1968) Nature *218*, 234.
STOKER, M.G.P. (1972) Proc. Roy. Soc. B. *181*, 1.

STOKER, M.G.P. (1973) Nature *246*, 200.

STREET, H.E. (1975a) *In:* Laboratory Manual of Cell Biology, Hall, D. and Hawkins, S. (eds.) (English Universities Press, London) p. 222.

STREET, H.E. (1975b) Ibid, p. 225.

STREET, H.E. (1977) Plant Tissue and Cell Culture, 2nd ed. (Blackwell, Oxford).

STRICKLAND, S. and W.H. BEERS (1976) J. Biol. Chem. *251*, 5694.

STRICKLAND, S., E. REICH and M.I. SHERMAN (1976) Embryol. Exp. Morphol. *18*, 155.

STRICKLAND, S. and V. MAHDAVI (1978) Cell *15*, 393.

STUBBLEFIELD, E. (1964) *In:* Cytogenetics of Cells or Culture, Harris, R.J.C. (ed.) (Academic Press Inc., New York) p. 223.

STUBBLEFIELD, E. (1973) Int. Rev. Cytol. *35*, 1.

STUBBLEFIELD, E. and G.C. MUELLER (1962) Cancer Res. *22*, 1091.

STUBBLEFIELD, E. and G.C. MUELLER (1965) Biochem. Biophys. Res. Comm. *20*, 535.

STUBBLEFIELD, E., R. KLEVECZ and L. DEAVEN (1967) J. Cell Physiol. *69*, 345.

STUBBLEFIELD, E. and C.M. DENNIS (1976) J. Theoret. Biol. *61*, 171.

STUDZINSKI, G.P. and W.C. LAMBERT (1969) J. Cell Physiol. *73*, 109.

TALAVERA, A. and C. BASILICO (1977) J. Cell Physiol. *92*, 425.

TAYLOR, C.R. (1978) Arch. Pathol. Lab. Med. *102*, 113.

TAYLOR, E.W. (1965) J. Cell Biol. *25* (1, part 2), 145.

TAYLOR, J.M. and C.P. STANNERS (1967) Biochim. Biophys. Acta *138*, 627.

TEGTMEYER, P. (1972) J. Virol. *10*, 599.

TEMIN, H.M. (1970) J. Cell Physiol. *78*, 161.

TERASIMA, T. and L.J. TOLMACH (1961) Nature *190*, 1210.

TERASIMA, T. and L.J. TOLMACH (1963) Exp. Cell Res. *30*, 344.

THOMPSON, E.B., G.M. TOMKINS and J.F. CURRAN (1966) Proc. Nat. Acad. Sci. U.S.A. *56*, 296.

THOMPSON, L.H., R. MANKOVITZ, R.M. BAKER, J.E. TILL, L. SIMINOVITCH and G.F. WHITMORE (1970) Proc. Nat. Acad. Sci. U.S.A. *66*, 377.

THOMPSON, L.H. and R.M. BAKER (1973) *In:* Methods in Cell Biology, Vol. 6. Prescott, D.M. (ed.) (Academic Press, Inc., New York) p. 210.

TOBEY, R.A. (1973) Methods in Cell Biology, Vol. 6, Prescott D.M. (ed.) (Academic Press Inc., New York) p. 67.

TOBEY, R.A., D.F. PETERSEN, E.C. ANDERSON and T.T. PUCK (1966) Biophys. J. *6*, 567.

TOBEY, R.A. and K.D. LEY (1970) J. Cell Biol. *46*, 151.

TOBEY, R.A. and K.D. LEY (1971) Cancer Res. *31*, 46.

TOBEY, R.A., D.F. PETERSEN and E.P. ANDERSON (1971) *In:* The Cell Cycle and Cancer, Baserga R. (ed.) (Marcel Dekker Inc., New York) p. 309.

TODARO, G.J. and H. GREEN (1963) J. Cell Biol. *17*, 299.

TODARO, G.J., G.K. LAZAR and H. GREEN (1965) J. Cell Comp. Physiol. *66*, 325.

TOLMACH, L.J. and P.I. MARCUS (1960) Exp. Cell Res. *20*, 350.

TOOZE, J. (1973) Molecular Biology of Tumor Viruses (Cold Spring Harbor Labs.).

TORMEY, D.C. and G.C. MUELLER (1965) Blood 26, 569.

TURNBULL, J.F. and R.L.P. ADAMS (1975) Nucl. Acids Res. 3, 677.

TURNER, M.K., R. ABRAMS and I. LIEBERMAN (1966) J. Biol. Chem. 241, 5777.

TURNER, M.K., R. ABRAMS and I. LIEBERMAN (1968) J. Biol. Chem. 243, 3725.

VAGO, C. (1971) Invertebrate Tissue Culture, Vols. I and II (Academic Press Inc., New York).

VAN DILLA, M.A., M.J. FULWYLER and I.U. BOONE (1967) Proc. Soc. Exp. Biol. Med. 125, 367.

VAN DILLA, M.A., T.T. TRUJILLO, P.F. MULLANEY and J.R. COULTER (1969) Science 163, 1213.

VAUGHAN, J.L. (1971) In: Invertebrate Tissue Culture, Vol. 1, Vago, C. (ed.) (Academic Press Inc., New York) p. 3.

VIGIER, P. (1970) 2nd Int. Symp. Tumour Viruses, 1969, p. 205.

VOGEL, A., E. RAINES, B. KARIYA, M.-J. RIVEST and R. ROSS (1978) Proc. Nat. Acad. Sci. U.S.A. 75, 2810.

WALTERS, R.A., R.A. TOBEY and R.L. RATLIFFE (1973) Biochim. Biophys. Acta 319, 336.

WALTHER, B., B. RAUSCH and S. ROSEMAN (1976) J. Cell Biol. 70, 70a.

WANG, H.C. and S. FEDEROFF (1972) Nature New Biol. 235, 52.

WANG, H.C. and S. FEDEROFF (1973) In: Tissue Culture, Methods and Applications, Kruse, P.F. and Patterson, M.K. (eds.) (Academic Press Inc., New York) p. 782.

WARMSLEY, A.M.H. and C.A. PASTERNAK (1970) Biochem. J. 119, 493.

WASLEY, G.D. and J.W. MAY (1970) Animal Cell Culture Methods (Blackwell, Oxford).

WATKINS, J.F. (1971) Methods of Virology, Vol. 5, Maramorosch, K. and Koprowski, H. (eds.) (Academic Press, Inc., New York) p. 1.

WEIGERT, M., L. GATMAITAN, E. LOH, J. SHILLING and L. HOOD (1978) Nature 276, 785.

WEINTRAUB, H. (1976) Cell 9, 419.

WEISS, M.C. and H. GREEN (1967) Proc. Nat. Acad. Sci. U.S.A. 58, 1104.

WESTALL, F.C., V.A. LENNON and D. GOSPODAROWICZ (1978) Proc. Nat. Acad. Sci. U.S.A. 75, 4675.

WESTERMARK, B. and A. WASTESON (1976) Exp. Cell Res. 98, 170.

VAN WEZEL, A.L. (1973) In: Tissue Culture, Methods and Applications, Kruse P.F. and Patterson, M.K. (eds.) (Academic Press Inc., New York) p. 372.

WHITAKER, A.M. (1972) Tissue and Cell Culture (Bailliere Tindall, London).

WHITE, M. and R. EASON (1973) Nature New Biol. 241, 46.

WHITMORE, G.F. and S. GULYAS (1966) Science 151, 691.

WIBLIN, C.N. and I.A. MACPHERSON (1972) Int. J. Cancer 10, 296.

WIGZELL, H. (1965) Transplantation 3, 423.

WILDY, P. (1971) Monogr. Virol. 5, 1.

WILLECKE, K., R. LANGE, A. KRÜGER and T. REBER (1976a) Proc. Nat. Acad. Sci.

U.S.A. *73*, 1274.

WILLECKE, K., P.J. DAVIES and T. REBER (1976b) Cytogen. and Cell Gen. *16*, 405.

WILLIAMS, A.F., G. GALFRE and C. MILSTEIN (1977) Cell *12*, 663.

WILLIAMSON, J.D. and P. COX (1968) J. Gen. Virol. *2*, 309.

WILLMER, E.N. (1965) Cells and Tissues in Culture, Vols. 1, 2 and 3 (Academic Press Inc., New York).

WITTES, R.E. and W.R. KIDWELL (1973) J. Mol. Biol. *78*, 473.

WYATT, G.R. and G.F. KALF (1957) J. Gen. Physiol. *40*, 833.

XEROS, N. (1962) Nature *194*, 682.

YAFFE, D. (1968) Proc. Nat. Acad. Sci. U.S.A. *61*, 477.

YAFFE, D. (1973) *In:* Tissue Culture, Methods and Applications, Kruse, P.F. and Patterson, M.K. (eds.) (Academic Press Inc., New York) p. 106.

YAMADA, K.M. and K. OLDEN (1978) Nature *275*, 179.

YAMANE, I., Y. MATSUYA and K. JIMBO (1968) Proc. Soc. Exp. Biol. Med. *127*, 335.

YUNKER, C.E., J.L. VAUGHN and J. CORY (1976) Science *155*, 1565.

ZAIN, B.S. (1971) Ph.D. Thesis, University of Glasgow.

ZAIN, B.S., R.L.P. ADAMS and R.C. IMRIE (1973) Cancer Res. *33*, 40.

Subject index